PRIMATES ON PRIMATES

PRIMATES ON PRIMATES

Approaches to the Analysis of Nonhuman Primate Social Behavior

Edited by

Duane D. Quiatt

University of Colorado
Boulder, Colorado

BURGESS PUBLISHING COMPANY   •   MINNEAPOLIS, MINNESOTA

Copyright © 1972 by Burgess Publishing Company
Printed in the United States of America
Library of Congress Catalog Card Number 72-88900
SBN 8087-1701-4

1   2   3   4   5   6   7   8   9   0

Contents

DUANE QUIATT   INTRODUCTION   vii

ONE   1

MARSHALL D. SAHLINS   The Social Life of Monkeys, Apes and Primitive Man   3

TWO   19

LIONEL TIGER AND ROBIN FOX   The Zoological Perspective in Social Science   21

THREE   29

VERNON REYNOLDS   Open Groups in Hominid Evolution   31

FOUR   45

CLAIRE RUSSELL AND W.M.S. RUSSELL   Primate Male Behaviour and Its Human Analogues   47

FIVE   59

DAVID A. HAMBURG   Evolution of Emotional Responses: Evidence from Recent Research   61

SIX   73

JOHN HURRELL CROOK   Social Organization and the Environment: Aspects of Contemporary Social Ethology   75

SEVEN   95

J. STEPHEN GARTLAN   Structure and Function in Primate Society   97

EIGHT   123

WILLIAM H. MASON   Scope and Potential of Primate Research   125

APPENDIX   135

Resources for the Study of Nonhuman Primate Behaviour   137

Introduction

BY DUANE QUIATT

The general theoretical papers reprinted here are intended to introduce the reader to the *study* of primate behavior, not the *nature* of primate behavior. Ideally, substantive description of the behavior of a number of species of contemporary nonhuman primates in field and laboratory settings should have been included to provide an immediate basis for assessing the validity of the insights (or, as it may be, the dubiousness of some of the propositions) set forth. However, such primary reports are readily available in most university libraries, and since the cost of including even a bare minimum would have been prohibitive I can only urge readers to consult as many as may be practical of those cited in authors' bibliographies.[1]

The new wave of primate studies has been interdisciplinary, with researchers in primate behavior operating from a number of biological, sociological, psychological, and anthropological perspectives, and the nonspecialist may find himself somewhat at a loss, at first, as to what to make of it all. However, a unifying objective of much primate research has been to frame questions concerning the nature and origins of human behavior, and the overall approach has been evolutionary and comparative. Such an approach may begin, of course, from any of a number of varying theoretical and methodological bases, and my aim in bringing together the authors represented here has been to present the student and interested reader with a distributed sample of that variety. Much of the literature of the last decade, the period spanned by this collection, falls along a polar continuum as far as basic assumptions go concerning the evolution of primate be-

---

[1] But see also the appendix to this collection: Resources for The Study of Nonhuman Primate Behavior.

havior. The ends are, to my mind, admirably defined by Crook in the introduction to his paper on Social Organization and the Environment. On the one end the emphasis is on individual behavior and fixed action patterns as expressed in diadic interaction: "The main fields of investigation continue to be motivation analysis, developmental studies and the evolution of species specific behavior." On the other end are studies which concentrate on the group as a whole (or, indeed, on larger populations) and which reflect to greater or lesser degree "a rapidly growing interest in the relations between ecology, population dynamics and social behavior." In these latter, the emphasis is on the group not just as a collection of interacting individuals but as "the social environment within which they live and to which they are adapted." Most of the papers that follow fall somewhere along this continuum. Some take a more or less traditional ethological approach in their discussion of the adaptive and evolutionary significance of items of individual behavior presumably under genetic control; others focus on group dynamics as these are affected by demographic, social behavioral, and ecological considerations. All, however, deal at one remove or another with inferences drawn from the study of nonhuman primates and applied to reconstruction of our hominid past.

Which of those inferences are both useful and valid I leave to the reader to decide on the basis of the papers themselves and the context which they provide one another. One aspect of that context, it may be observed, is historical. Most of the papers represent alternative general synthesizing approaches to data as they were gathered over the last decade. The views presented here were developed as the data were being gathered. As alternative views these haven't changed much, however, despite new information, and the papers that follow should make up for those accustomed to questioning what they read a compact introduction to general theory concerning primate social behavior.

What this collection is intended to provide, then, is a spectrum of ideas, a range of interpretative essays. All of them involve some measure of speculation, and I have refrained, in my accompanying remarks, from grading orders and levels of speculative hypotheses from any doctrinaire standpoint. I would like only to remind the reader that those who extrapolate from the behavior of the living nonhuman primates to either the behavior of early hominids or to that of contemporary men do so on the basis of assumptions which are often unspecified. The reader who is not well grounded in the theory of biological evolution and at least some of the data of comparative animal behavior would do well to reflect and reflect again before affirming the validity of any given interpretation of primate behavior in terms of its adaptive significance and its relevance to evolutionary reconstruction.

One

Marshall Sahlins's paper may be thought of as providing a dateline for modern research into the behavior of nonhuman primates. At the time of its writing, available information on primate social behavior in natural settings did not afford "comfortable support for weighty generalizations," and Sahlins warns his readers at the outset that his interpretations of that behavior are to be taken as provisional. In particular, his view that nonhuman primate society can be characterized as territorial, with intrasocietal relations strongly influenced by sex-linked aggressive dominance, would require considerable modification today. How-

1

ever, this view, derived from Zuckerman and always popular in the writings which have most influenced anthropologists, dies hard — just how hard will be evident to anyone familiar with the spate of popularized ethology which has lately appeared.

Sahlins sees a generic biological continuity but a specific behavioral discontinuity between nonhuman primate society and human cultural society. He lists toward the end of this paper five "significant advances in the early evolution of cultural society defined by comparison of primate and primitive human social groups." To problem-oriented primatologists, interested in the continuities rather than the discontinuities between nonhuman and human primate behavior, the list might be one of areas for fruitful investigation.

The Social Life of Monkeys, Apes and Primitive Man

BY MARSHALL D. SAHLINS[1]

INTRODUCTION

This study compares societies of infrahuman primates with the most rudimentary of documented human social systems. The objectives are to describe general trends in primate social organization leading to human society, and to delineate the major advances of the latter, cultural society, over precultural society.

For comparative materials, we rely on field studies of monkey and ape social behavior, supplementary observations of these animals in captivity, and ethnographic accounts of simple hunters and gatherers. The quality and quantity of published studies of free-ranging subhuman primate societies do not provide comfortable support for weighty generalizations. Aside from the anecdotal literature, we have only Carpenter's accounts of spider monkeys, rhesus, howling monkeys, and gibbons; Nissen's chimpanzee material; Zuckerman's observation of baboons (perhaps biased by his captivity studies); the as yet incomplete reports on the Japanese monkey; and some peripheral notes on the African red-tailed monkey by Haddow, and on the gorilla by Schwab.[2] *Considering this, our interpretations of subhuman primate social behavior are entirely provisional.* The data on primitive food gatherers are more abundant. We include in our comparison the following primitive societies: Australian Aborigines, Tasmanians, Semang, Andamanese, Philippine and Congo pygmies, Bushmen, Eskimo, Great Basin Shoshoni, Naskapi, Ona and

Reprinted from *Human Biology*, Vol. 31.1, February 1959, pp. 54-73, by permission of the Wayne State University Press. Copyright 1959 by Wayne State University Press.

[1] I thank Professors M.H. Fried, F.P. Thieme, S.L. Washburn, E.R. Service, J.N. Spuhler and L.A. White for their useful criticisms of an earlier version of this paper. I am especially indebted to White's articles touching subjects dealt with here (White, 1949, Chapters II, III, VI, XI). Needless to say, the above named do not necessarily agree with all statements contained herein.
[2] The following provide the field data on subhuman primates used in this paper: Carpenter, 1934; 1935; 1940; 1942c; Collias and Southwick, 1952; Haddow, 1952; Hooton, 1942 (report of Schwab's observations); Imanishi, 1957; Nissen, 1931; and Zuckerman, 1932. Unless quoting or making special points, these works will not be specifically cited hereafter.

Yahgan.[3] It is assumed that these societies parallel early cultural society in general features. This is simply an assumption of order and regularity. The technologies and low productivity of modern hunters and gatherers resemble the archaeologically revealed productive systems of early cultures. Granting that a cultural social system is functionally related to its productive system, it follows that early human society resembles rudimentary, modern human society. This reasoning is supported by the large degree of social similarity among the present hunters and gatherers themselves, despite the fact that some of them are as historically distant from each other, as separated in contact and connection, as the Paleolithic is separated from modern times. Further, simply because many food gatherers have been driven into marginal areas, they are not thereby disqualified from consideration. There still remain strong social resemblances between marginal peoples, such as Bushmen, Ona, and Eskimo, and those found in isolated, but otherwise not ecologically marginal areas, such as many Australian groups and the Andaman Islanders.

A comparison of subhuman primate and primitive society must recognize the qualitative difference between the two (see White, 1949: Chapter III). Human society is cultural society; the organization of organisms is governed by culture traits. The social life of subhuman primates is governed by anatomy and physiology. Variations in human society are independent of, and are not expressions of, biological variations of the organism. Variations in primate society are direct expressions and concomitants of biological variation. Nissen writes: ". . . with one notable exception the phylogenetic course of behavioral development has been gradual . . . it has been a continuous affair, proceeding by quantitative rather than qualitative changes. The one exception is that which marks the transition from the highest nonhuman primates to man . . . At this point a new 'dimension' or mode of development emerges: culture" (Nissen, 1951:426).

It follows that assertions of specific phylogenetic continuities from anthropoid to primitive society must be summarily rejected, such as Yerkes' suggestion that delousing among primitives is a genetic survival of primate grooming, an activity which, Yerkes writes, also led to: "tonsorial artistry, nursing, surgery, and other social services of man" (Yerkes, 1933: 12). In the same vein is Kempf's identification of the presenting behavior of a subordinate rhesus monkey with human prostitution (Kempf, 1917, cf. Miller, 1931). Furthermore, the terminology of cultural social and political organization should be disavowed when describing infrahuman primate society. The cultural anthropologist justifiably shudders when he reads of "clans," "com-

---

[3] The following provide the field data on hunters and gatherers used in this paper: Birket-Smith, 1936; Bleek, 1928; Boas, 1888; Cooper, 1946a; 1946b; Elkin, 1954; Forde, 1934; Gusinde, 1955; Leacock, 1954; 1955; Lothrop, 1928; Man, 1885; Putnam, 1953; Radcliffe-Brown, 1930-1931; 1948; Rink, 1875; Roth, 1890; Schapera, 1930; Schebesta, 1933; n.d.; Sharp, 1934-1935; Spencer and Gillen, 1927; Steward, 1938; Vanoverbergh, 1925; Warner, 1937; and Weyer, 1932. Unless quoting or making special points, these works will not be specifically cited hereafter.

munism," and "socialism," among howler monkeys, or, "despots," "tyrants," "absolutism," and "slavery," among baboons.

The determinants of sociability are different in cultural and precultural society. We find useful Zuckerman's contention that social organization in general is based upon, "three main lines of behavior — search for food, search for mates, avoidance of enemies" (Zuckerman, 1932: 17). Of these factors the sexual one appears to be primary in the genesis of subhuman primate society: "The main factor that determines social groupings in subhuman primates is sexual attraction." *(Ibid:* 31). "The emergence of this feature [i.e., of continuous sex activity] into prominence in their behavior has created primate society" (Chance and Mead, 1953: 415). It was the development of the physiological capacity to mate during much of, if not throughout, the menstrual cycle, and at all seasons, that impelled the formation of year round heterosexual groups among monkeys and apes.[4] Within the primate order, a new level of social integration emerges, one that surpasses that of other mammals whose mating periods, and hence heterosexual groupings, are very limited in duration and by season. Certainly, defense against predation is also a determinant of subhuman primate sociability (Chance, 1955), but it, and the search for food, appear to be secondary to sex. The influence of sexual attraction in promoting solidarity among subhuman primates has been noted in the field. Carpenter has observed of the howlers: "With repetition of the reproductive cycle in the female and with uninterrupted breeding throughout the year, the process of group integration through sexual behavior is repeatedly operative, establishing and reinforcing intersexual social bonds" (Carpenter, 1934: 95; for a common alternate view, see Yerkes and Yerkes, 1935: 979).

Sexual attraction remains a determinant of human sociability. But it has become subordinated to the search for food, to economics. A most significant advance of early cultural society was the strict repression and canalization of sex, through the incest tabu, in favor of the expansion of kinship, and thus mutual aid relations (Malinowski, 1931; Tylor, 1888; White, 1949: Chapter XI). Primate sexuality is utilized in human society to *reinforce* bonds of economic and/to a lesser extent, defensive alliance. "All marriage schemes are largely devices to check and regulate promiscuous behavior in the interest of human economic schemes" (Miller, 1931: 402). This is not to underestimate the importance of primate sexuality in determining certain *general* characteristics of human society. If culture had not developed in the primate line, but instead among creatures practicing external fertilization, marriage and rules of exogamy and endogamy would not be means of establishing cohesive groups in cultural society. But, in the transition from subhuman to human society, cooperation in subsistence activities became the dominant cause of solidarity, avoidance of enemies a secondary cause, while sex became simply a facilitating mechanism.

[4] The Japanese monkey, *Macaca fuscata*, northernmost of all subhuman primates, is reported to have a breeding season (Imanishi, 1957).

These propositions are best documented by detailed consideration of primate and primitive society, wherein the differences in causes of sociability will be seen to pervade the comparison, and to turn generic continuity into specific discontinuity.

SUBHUMAN PRIMATE AND HUMAN PRIMITIVE SOCIETIES

Territoriality is one of a number of common features of subhuman primate and primitive social behavior. It is also general among lower vertebrates — perhaps it is a universal characteristic of society. Territoriality arises from competition over living conditions, and has the selective advantage of distributing the species in its habitat so as to maintain population density at or below its optimum (Allee, *et al.*, 1949: 411f; Bartholomew and Birdsell, 1953: 485).

Territorial relations among groups of subhuman primates of the same species are generally exclusive. Except for the few animals that are driven out of one group and may attach themselves to another, primate societies are usually semi-closed societies (Carpenter, 1942a: 187). Each horde has a focus of favorite feeding and resting places to which it is often deflected by contact with other groups. Contact between groups of the same species at territorial borders is generally competitive and antagonistic, sometimes violently so. Subhuman primate groups apparently have little tendency to federation. They *"do not have supergroup social mechanisms* . . . Kinship relations are not operative and inbreeding is the rule rather than the exception . . ." (Carpenter, 1954: 98). Carpenter finds the origin of intergroup cooperation characteristic of primitive tribalism, "difficult to trace in subhuman primates which I have studied" (1940: 163).

Primate territorial relations are altered by the development of culture in the human species. Territoriality among hunters and gatherers is never exclusive, and group membership is apt to shift and change according to the variability of food resources in space and time. Savage society is open, and corresponding to ecological variations, there are degrees of openness: 1. Where food resources are evenly distributed and tend to be constant year to year, territory is clearly demarcated and stable, and, considering the nucleus of males, so is group membership. The Ona and most Australian groups are representative of this condition. Stability of membership is effected through customary rules of patrilocal residence, which is usually coupled with local exogamy. Patrilocal residence confers the advantages of continuity of occupation for hunters in areas with which they are familiar. 2. Where food resources are evenly distributed during some seasons, and variably abundant during others, exclusiveness of territory and membership are only seasonal. Local groups of Central Kalahari Bushmen, for example, remain fixed in territories focused around water holes during the dry season; whereas, in the rainy season, such groups mingle and hunt together. Similarly the Semang have seasonal territories fixed by the distribution of the durian tree; after the harvest, territories and exclusive groups dissolve, to be reconstituted at the next durian season. Under these conditions, patrilocal residence and local exogamy are preferred, but not strict, rules, and local endogamy and matrilocality occur. The Andaman Islanders with relatively fixed territories, a tendency toward local exogamy, and no residence rule, and the Yahgan, who tend to be locally exogamous, patrilocal, and territorially exclusive, may also fall into this ecological type. However, data available for classification of these groups are inadequate. 3. Finally, there are food gatherers among whom rules of territoriality are *de facto* nonexistent, and local group composition highly variable. These occupy areas where food resources vary in local abundance seasonally and annually. Families coalesce and separate *ad hoc* corresponding to accessibility of supplies. Postmarital residence may be in the group of either spouse, and band aggregates are agamous — there are no rules. The Great Basin Shoshone, the Eskimo, and the pre-fur trade Naskapi fall into this category.[5]

Territoriality among hunters and gatherers is sometimes maintained by conflict. There appears to be a direct relationship between intensity of intergroup feud and exclusiveness of territory and membership. Thus trespass is strongly resented and interband feuds are relatively frequent among Australians and the Ona; whereas, where territoriality is weak, as among Eskimos and Shoshoni, the concept of trespass is naturally poorly developed, and fighting consists of squabbles between particular families. However, in all cases, the outcome of trespass depends on the previous relations between neighbors, and these are usually *friendly*. Even among the Australians, adjacent groups would be allowed to hunt in a band's territory if in need; there is "no constant state of enmity" between neighbors (Spencer and Gillen, 1927, i: 58). Everywhere, no matter how strict the rules of territory, constant visiting between bands prevents the development of closed groups. And everywhere exclusiveness is easily broken down if there is some food windfall, or if food is differentially abundant in adjacent locales.

Hunters and gatherers live in relatively open groups between which relations are usually friendly; infrahuman primates of the same species live in relatively closed groups between which relations are usually competitive. The invention of kinship and the incest tabu of cultural society are responsible for this difference. Through marrying out, friendly, cooperative relations are established between families. When exogamy can be extended to the local group, cooperation between bands is effected (Tylor, 1888; White, 1949; Chapter XI). Intermarrying Australian bands are described as, "*a family* of

[5] Philippine and Congo pygmies are presently of the first type: rigid territory, patrilocal, exogamous. However, both live in symbiotic, subservient relation to agricultural peoples; and, at least for the Congo Twides, territorial exclusiveness is clearly a function of boundaries between patron Bantu villages (Putnam, 1953; Schebesta, 1933; Gusinde would disagree, 1955:23). The Heikum Bushmen, also in symbiotic, subservient status to patrilineal Hottentot and Bantu groups, are another instance of the same thing. Unlike other Bushmen, the Heikum live in well-defined territories and practice strict patrilocality and local exogamy (Schapera, 1930: 34-35, 38, 83, 94f). Leacock's (1955) Naskapi studies suggest that bilocal, nonterritorial, unstable bands commonly become so formalized under outside influences of this general sort.

countries bound together by those sentiments which function between members of a family and its near relations" (Elkin, 1954: 81; emphasis ours). It is the kinship ethic of mutual aid that permits populations of hunters and gatherers to shift about according to the distribution of resources. Kinship is thus selectively advantageous in a zoological sense; it permits primitives to adjust to more variable habitats than subhuman primates (see Carpenter, 1955: 93). Moreover, the kinship relations between groups, and the ceremonies and exchanges of goods that frequently accompany interband meetings, give rise to a further social development: tribalism. Common custom, common dialect, a name and a feeling of unity are created among otherwise independent groups. The stage for further political evolution is thereby set.

We turn now to the internal organization of subhuman primate and primitive human societies.

The subhuman primate horde varies in size from an average of four animals among the gibbon to several hundred baboons (Carpenter, 1942a). Group size is not correlated with suborder differences, except that great ape hordes are generally at the lower end of the primate range; orangutan groups are apparently as small as gibbons' and chimpanzees average 8.5 per group. The horde may remain together at all times, or may disperse, during daytime feeding, into segments of various constitution — mating groups, female packs, male packs — concentrating at night resting places. Howling monkeys typically travel together; spider monkey and baboon groups are instances of segmented hordes.

With the exception of the gibbon, the primate horde characteristically contains more adult females than adult males. In observed wild groups the ratio ranges from nearly 3:1 for howlers, to 1:1 for gibbons. The ratio for spider monkeys, 1.6 females per male, is probably near the average for the primates. The usual inequality apparently reflects the degree of dominance and competition among males for female sexual partners.[6] There are almost always unmated males, either peripherally attached to heterosexual groups, or existing outside the horde.

We have argued that sexual attraction is the primary cause of subhuman primate sociability. Indeed, in many cases the entire horde is a single reproduction unit or mate group. But there are significant species variations in the constitution of hordes and mate groups. There appears to be a progressive development within the primate order from promiscuous relations within the group to the establishment of exclusive sex partnerships, one of which comprises the nucleus of a horde. Since the young remain attached to adult females throughout, the highest forms of primate mate groups resemble the elementary human family in composition (Carpenter, 1942a: 186; Chance and Mead, 1953: 418; Yerkes and Yerkes, 1929: 566). The emergence of exclusive, independent mate groups takes the following steps: 1. Among New World howler and spider monkeys observed in free-ranging conditions, the

[6] Washburn has suggested (personal communication) that the unequal ratio may in good part be due to the faster maturation rate for females.

only stable relations within a horde are between females and their young. Only when females are in the oestrus period of the menstrual cycle do they leave the female-offspring pack and become attached to specific males, and then not exclusively, but to several in succession. The mated pair is a temporary, non-exclusive unit. 2. Old World monkeys develop more permanent sex partnerships. Rhesus shows the trend toward exclusiveness. Again in rhesus, the female-young pack is a separate unit, and sex partnerships are only temporarily established while a female is in heat. However, for every female, the succession order among her male partners corresponds to their dominance position. Therefore, dominant males have all females in oestrus, while subordinate males are excluded from some when there are not enough to go around. The mate group of the Japanese monkey, in the same genus as rhesus, is very similar. 3. The baboon mate group is a simple extrapolation from rhesus. The steps involved are: the exclusion of subordinate males from sexual relations with females, and the development of constant association between a dominant male, females and young. The baboon mate group is a permanent, exclusive relationship between a dominant male and his several females, the young following their mothers. Subordinate males may remain attached to the group on its fringes, or form unisexual bands. The baboon horde consists of several such mate groups and male bands, each relatively independent. 4. The ape horde tends to be composed of a single, independent mate group of the baboon type. In the gibbon, this consists of one male, one female plus young. The evidence from the other anthropoids is not conclusive; however, gorilla, orangutan and chimpanzee hordes apparently consist of a single mate group of one male, two or more females and their young, and perhaps subordinate males.

The emergence of exclusive, permanent mate groups among higher primates is explained by the progressive emancipation of sexual behavior from hormonal control running through the order (Beach, 1947). In monkeys, copulation outside the female's fertile period is relatively rare. In apes — more commonly so, in man — sex is freed from hormonal regulation, being subject instead to cortical and social control.[7] The oft-made alternative assertion that the development of the family is due to increasing duration of infant dependency is not supportable. With one minor exception, subhuman primate males are never significantly involved in rearing the young, save in retrieving the fallen, and indirectly as group defenders. The exceptional case of the male Japanese monkey that rears the older infant if a female has two in succession is not significant here, since there is no family-type group involved. This behavior was observed only in one of a number of hordes of Japanese monkeys, and is entirely unique among subhuman primates. Baboon and rhesus males are known to have killed young in the course of sexual attacks upon their mothers. A long dependency period cements mother-offspring

---

[7] This is a crucial physiological change for a number of reasons, not the least of which is that it makes intelligible how, when culture developed, primate sexuality was so radically subordinated to other ends.

relations but not necessarily father-mother-offspring relations. Only when there is an economic division of labor by sex can infant dependency produce this effect (Dole, n.d.). In subhuman primates there is no sexual division of economic labor.

Social relations within the subhuman primate horde vary according to the age, sex and dominance statuses of the interacting animals. Leaving aside dominance for a moment, most social interaction can be adequately described by utilizing three elemental status categories: adult male, adult female, and young (Carpenter, 1942a: 180). Interaction of animals of these categories produce 6 "qualitatively distinct" social relations: male-male; female-female; male-female; male-young; female-young; young-young (Carpenter, 1940: 126). The content of most of these relations can be inferred from the preceding and succeeding discussion. It will be seen that age and sex difference remain important social distinctions among primitive hunters and gatherers.

Dominance statuses are found among all known monkeys and apes, as well as many lower vertebrates (Carpenter, 1942a; Maslow, 1936a). Dominance is established by competition for mates, for food, for position in progression, and the like − ". . . in every known typical grouping of monkeys and apes there is persisting competition for priority rights to incentives" (Carpenter, 1942a: 192). Conflict is often particularly prominent among males over females in oestrus. Males are usually dominant over females. [This last is subject to exception when females are in heat (Crawford, 1940; Nowlis, 1942; Yerkes, 1940). Dominance has been experimentally raised in primates and other vertebrates by injections of male sex hormones (Clark and Birch, 1945; Birch and Clark, 1946).] Dominance status affects behavior in every aspect of social life: play, feeding, sex, grooming, competition between groups, and it even determines the spatial relations of animals within the horde (Carpenter, 1942a; Chance and Mead, 1953; Maslow, 1936a; 1936b; Maslow and Flanzbaum, 1936; Nissen, 1951). Dominance among paired animals is easily determined experimentally by introducing a series of food pellets to which there is limited access, and noting which animal consistently appropriates them (Maslow, 1936a; Nissen, 1951: 447; Nowlis, 1941a; 1941b). Nowlis' food appropriation experiments, performed with differentially satiated animals, show that dominance is a social behavior arising from conflict − not a simple, independent drive for "prestige," as is sometimes claimed (Nowlis, 1941b; 1942).

Maslow contends that the quality of dominance varies among New and Old World monkeys and the apes, and that variation in dominance quality is correlated with differences in social organization (Maslow, 1940). Platyrrhines, according to Maslow, show the greatest indifference in social relations. Dominance is "tenuous," "non-contactual," frequently not expressed, and often ascertained in the laboratory only with difficulty. In contrast, "Catarrhine dominance is rough, brutal and aggressive; it is of the nature of a powerful, persistent, selfish urge that expresses itself in ferocious bullying, fighting and sexual aggression" (Maslow, 1940: 316). Weak and sick animals are attacked; in competition over food, a subordinate animal would starve. Chimpanzees, however, show "friendly dominance." Dominant animals protect subordinates, never attack them except in the form of rough play. Crawford (1942) in one laboratory experiment noted that only 0.4% of chimpanzee social behavior could be described as aggressive, and other field and captivity data generally bear out Maslow's hypothesis of suborder differences in dominance quality (cf. Carpenter, 1934; 1937; 1942a; Collias and Southwick, 1952; Gillman, 1939; Harlow and Yudin, 1933; Maslow, 1936b; Warden and Galt, 1943; Yerkes and Yerkes, 1929; 1935; Zuckerman, 1932; for an exception, see Haddow, 1952).

There appear to be correlated differences in the grooming behavior of the suborders. Yerkes advances the notion that social, as opposed to self-grooming, increases in frequency from prosimian through anthropoid ape (Yerkes, 1933). Grooming serves a biological function in removing parasites and the like, but this is nearly equally accomplished by self or social grooming. Therefore, social grooming takes on added significance as a "social service" in the higher primates. Moreover, grooming in higher primates is not only social, but reciprocal. Social grooming among wild gibbons has been noted to be frequent and reciprocal. By contrast, evidence from field and laboratory indicates that social grooming is comparatively infrequent among New World monkeys. It is difficult to say from present evidence that social and reciprocal grooming increase from Old World monkeys to anthropoids, but on the whole, Yerkes' assertions are supportable (cf. Carpenter, 1935; Crawford, 1942; Maslow and Flanzbaum, 1936; Warden and Galt, 1943).

The emergence of reciprocal social behavior and the progressive tempering of dominance relations are significant trends in the primate line — trends which, we shall note, are continued in primitive society. On the other hand, economic teamwork and mutual aid are nearly zero among subhuman primates, including anthropoids. Spontaneous cooperation — as opposed to one animal helping another — has not been observed among them. Chimpanzees have been trained to solve problems cooperatively, but fail to do so without tuition (Crawford, 1937). Monkeys apparently cannot even be taught to cooperate (Warden and Galt, 1943). Spontaneous teamwork presupposes symboling: "Teamwork makes intellectual demands of the same order as those made by language. Psychologically, it may, in fact, be difficult to distinguish between the two" (Hebb and Thompson, 1954: 540). Nissen and Crawford's elaborate experiment showed that chimpanzees share food pellets and tokens, although sharing was much less frequent than not, and was evidently non-reciprocal (Nissen and Crawford, 1936). However, in a similar experiment using animals tested for dominance, Nowlis observed that every case of sharing (in 480 trials there were 80 instances of sharing and, in these, one-tenth of the food available was shared) was from subordinate to dominant; dominants never gave food to subordinates (Nowlis, 1941). Therefore, "food sharing" here is a function of previous dominance competition and actually indicates monopolization, not pooling, of a limited supply (Nissen, 1951: 447).

We now consider the social system of bands of hunters and gatherers. The usual band contains 20 to 50 people (Steward, 1955: 146), but during poor seasons it may fragment into small family groups. We have already looked at band structure. Corresponding to ecological conditions, bands range in composition from an enlarged patrilocal family to a congeries of variably related nuclear families. Almost all members will be kinsmen, and the etiquette of kinship behavior regulates social life. In Australia, for example, kinship: ". . . regulates more or less definitely the behavior of an individual to every person with whom he has any social dealings whatsoever" (Radcliffe-Brown, 1930-1931: 43).

The family is the only social unit inside the band; where bands are unstable, it is the major form of social organization. Among primitives, the division of economic labor by sex is fundamental to the family and makes marriage an economic alliance — or, in Westermarck's terms, ". . . marriage is something more than a regulated sexual relation. It is an economic institution" (Westermarck, 1922, i: 26). The complementary economic roles of the sexes determine certain qualities of primitive marriage. First, it is a necessity for all adults; the unmated adult male of the subhuman primate horde rarely has a counterpart in primitive bands. Secondly, polygamy is usually economically impractical; monogamy is prevalent. Finally, as indicated, stable heterosexual relations are not simply determined by sexual attraction, but by economics. Sex is easily had in many hunting and gathering groups, both before and beside marriage, but such neither necessarily establishes the family nor destroys it. Legal rights to sexual privileges of spouses may be waived in favor of socio-economic advantages, as in wife lending. The very rules of exogamy and incest prevent the formation of the family on a basis of simple sexual attraction. Steward's statement of the economic basis of the Shoshoni marriage can be duplicated from accounts of other simple societies: "Marriage was an economic alliance in a very real sense . . . a union which brought into cooperation the complementary economic activities of the sexes — a person could not, in the interest of self-preservation, afford to remain long single . . . the role of exclusive sex privileges in matrimony seems to have been secondary" (Steward, 1938: 242). Compare Radcliffe-Brown on the Australians: "The family is based on the cooperation of man and wife, the former providing the flesh food and the latter the vegetable food . . . this economic aspect of the family is a most important one . . . I believe that in the minds of the natives themselves this aspect of marriage, i.e., its relation to subsistence, is of greatly more importance than the fact that man and wife are sexual partners" (Radcliffe-Brown, 1930-1931: 435).

In man, therefore, primate sexuality has been brought under cultural control; it has become, in part, a means to other ends. Another aspect of primitive marriage teaches the same lesson. Unlike primate unions, created and maintained in conflict, primitive marriage is a powerful factor in interfamilial and interband alliance. Again, as Steward writes of the Shoshoni, and one can find countless ethnographic paraphrases of this truism: "Marriage was more a contract between families than between individuals" (Steward,

1955: 118). Among hunters and gatherers, marriages are frequently arranged (or at least approved) by the families rather than the spouses. Considerations often pass between the groups, thus setting the pattern for future cooperation. Women may be exchanged between groups, or intermarriage between certain parties preferred and repeated, solidifying both the marriages and group relations. The alliance between families may be paramount to the extent that a marriage can survive the death of one of the partners, he or she being replaced through the levirate or sororate.

There is an outstanding implication of these characteristics of primitive marriage and the family. Given the division of labor by sex and the formation of domestic units through marriage, it follows that sharing food and other items, rather than being nonexistent, as among monkeys and apes, is a *sine qua non* of the human condition. Food sharing is an outstanding functional criterion of man. In the domestic economy of the family there is constant reciprocity and pooling of resources. And, at the same time that kinship is extended throughout the band of families, so are the principles of the domestic economy. Among all hunters and gatherers there is a constant give and take of vital goods through hospitality and gift exchange. Everywhere, generosity is a great social virtue. Also general is the custom of pooling large game among the entire band, either as a matter of course, or in times of scarcity. Where kinship is extended beyond the local group by interband marriage, so are reciprocity and mutual aid. Goods may pass over great distances by a series of kinship transactions. Trade is thus established. Hunters and gatherers are able to take mutual advantage of the exploitation of distant environments, a phenomenon without parallel in the primate order. This emphasis on generosity, on mutual aid, and the attribution of social prestige for generosity, stand in direct opposition to the tendency among primates to monopolize vital goods. Perhaps the elaborated emphasis on sharing among primitives is to be partially understood as a cultural means for overcoming primate tendencies in the opposite direction.

In the system of social status, there is a generic continuity between primate and primitive society. Qualitative social differences of sex and age that are marked in subhuman primate groups are major principles of status and role allocation among hunters and gatherers. The division of labor by sex is an example. So is the pervasive recognition of sex, seniority and generation in kinship behavior and terminology (e.g., see Radcliffe-Brown, 1948).

There is also some continuity in dominance status. Leadership falls to men among hunters and gatherers, although, what is possibly different from subhuman primates, it is especially the elders that are respected. There are, however, important qualitative differences in dominance relations among hunters and gatherers and subhuman primates. Each primitive band usually has an elderly headman. The respect accorded him and other elders is not due to their physical ability to appropriate a limited supply of desired objects. (They may be preferentially treated in communal food distribution, but this is another thing.) Quite the opposite from subhuman primates, a man must be generous to be respected. Prestige among hunters and gatherers can be estimated by noting who gives away the most — precisely the reverse of the test for dominance status among subhuman primates. The position of the head man rests primarily on superior knowledge of game movements, water and other resources, ritual and other things which govern social life. Thus Boas pointed out that there is a direct relation between the authority of Eskimo headmen of various groups and the distance and difficulties involved in traveling between winter and summer hunting grounds (Boas, 1888). But such knowledge alone cannot breed power. The leader of the band has no means to compel obedience. He is commonly described as ruling through "moral influence." A Congo pygmy leader bluntly told Schebesta, "There would be no point in his giving orders, as nobody would heed them." (Schebesta, 1933: 104). Steward comments that the title, "talker," given to a Shoshoni leader, "truly designates his most important function" (Steward, 1938: 247). The leader of the Central Eskimo camp is picturesquely referred to as *isumataq*, "he who thinks" (for others) (Birket-Smith, 1936: 148). In all bands of hunters and gatherers, the heads of the separate families exercise more control than the informal headmen over the whole. Compared to infrahuman primates, ranking hierarchies and dominance approach zero among hunters and gatherers. Yet, where all interact as kinsmen, and sharing a scanty food supply replaces conflict over it, this is expectable.

SUMMARY AND CONCLUSIONS

The transition from subhuman primate society to rudimentary cultural society was at the same time a process of generic continuity and of specific discontinuity. If culture had not developed among a species of primates, but among animals of different behavioral characteristics, then the forms and development of cultures would be basically different. The social behavior of primates is the foundation of some general features of human society. On the other hand, no specific trait of cultural society, even in its most rudimentary state is, in both form and functioning, a direct survival of some specific trait of primate social behavior. This discontinuity is due to the fact that subhuman primate society is a direct expression of the physiology of the species operating in a given environment; whereas cultural traits govern the social adaptation of the human primate. The development of culture did not simply give expression to man's primate nature, it replaced that nature as the direct determinant of social behavior, and in so doing, channeled it — at times repressed it completely. The most significant transformation effected by cultural society was the subordination of the search for mates — the primary determinant of subhuman primate sociability — to the search for food. In the process also, economic cooperation replaced competition, and kinship replaced conflict as the principal mechanism of organization.

What are the generic continuities? Territoriality is one. But similarities such as these are common to a wide variety of societies, including those of lower vertebrates. A more restricted continuity is the utilization of the powerful social functions of primate sexuality in human social organizations.

To repeat an earlier observation, it is only against the unique background of primate sexual behavior that one fully understands why marriage and marriage rules are general mechanisms for integrating cooperative human groups. Another generic survival of simian society is the allocation of social functions on the lines of sex and age among hunters and gatherers.

Of particular interest are social advances within the primate order, upon which cultural society directly elaborated. Here the cultural developments appear capstones to trends which had begun to unfold in precultural conditions. In the primate line the exclusive mate group appears to have developed out of the promiscuous horde. In the transformation to the human family, the anthropoid mate group was altered more in function than in form. A second primate advance is the development of reciprocal social services in grooming. Generalized to food sharing, reciprocity is basic to cultural society. Thirdly, there is the softening of dominance relations among higher primates. In primitive society, dominance or prestige is especially associated with service to the group. (See also Hallowell, 1956, on generic similarities.)

The most significant advances in the early evolution of cultural society can be deduced by comparison of primate and primitive groups. To us these advances appear to be: 1. the division of labor by sex and the establishment of the family on this basis; 2. the invention of kinship; 3. the incest prohibition and its extension through exogamy, thus extending kinship; 4. the overcoming of primitive competition over food in favor of sharing and cooperation; and 5. the abolition of other primate conflicts leading to the establishment of dominance hierarchies.

These 5 are complementary; nothing is said here of their order of appearance or relative significance. It is claimed that they are great triumphs of early culture. These developments are necessary social counterparts of the continuous tool activity which enabled man to become the dominant form of life.

REFERENCES

Allee, W.C.; Emerson, Alfred E.; Park, Orlando; Park, Thomas; and Schmidt, Karl P. 1949. Principles of Animal Ecology. Philadelphia and London: W.B. Saunders.

Bartholomew, George A., and Birdsell, Joseph B. 1953. Ecology and the protohominids. Am. Anthrop. 55:481-98.

Beach, Frank A. 1947. Evolutionary changes in the physiological control of mating behavior in mammals. Psychol. Rev. 54:297-315.

Birch, Herbert S., and Clark, George. 1946. Hormonal modification of social behavior (II). The effects of sex hormone administration in the social dominance status of the female-castrate chimpanzee. Psychosom. Med. 8:320-31.

Birket-Smith, Kaj. 1936. The Eskimos. London: Methuen.

Bleek, D.F. 1928. The Naron, A Bushman Tribe of the Central Kalahari. Cambridge: Cambridge University Press.

Boas, Franz. 1888. The Central Eskimo. Smithsonian Institution, Bureau of Ethnology, Annual Report 6:401-669.

Carpenter, C.R. 1934. A field study of the behavior and social relations of howling monkeys. Comp. psychol. Monogr. 10(2):1-168.

_____. 1935. Behavior of red spider monkeys in Panama. J. Mammal. 16:171-180.

———. 1937. An observational study of two captive mountain gorillas. Human Biol. 9:175-96.

———. 1940. A field study in Siam of the behavior and social relations of the gibbon (Hylobates lar). Comp. psychol. Monogr. 16(5):1-212.

———. 1942a. Societies of monkeys and apes. In Biological Symposia 8:177-204.

———. 1942b. Characteristics of social behavior in non-human primates. Trans. N.Y. Acad. Sci. 4:248-58.

———. 1942c. Sexual behavior of free ranging rhesus monkeys (Macaca mulatta) I and II. J. comp. Psychol. 33:113-42; 143-62.

———. 1954. Tentative generalization on the grouping behavior of non-human primates. Human Biol. 26:269-76. Reprinted in James A. Gavan, ed., 1955, The Non-Human Primates and Human Evolution. Detroit: Wayne University Press.

Chance, M.R.A., and Mead, A.P. 1953. Social behavior and primate evolution. Symposia of the Soc. for Experimental Biol. 7:395-439. New York: Academic Press.

Chance, M.R.A. 1955. The sociability of monkeys. Man 55(176):162-65.

Clarke, George, and Birch, Herbert. 1945. Hormonal modifications of social behavior 1. The effect of sex-hormone administration on the social status of a male-castrate chimpanzee. Psychosom. Med. 7:321-29.

Collias, Nicholas, and Southwick, Charles A. 1952. A field study of population density and social organization in howling monkeys. Proc. Am. Philos. Soc. 96:143-56.

Cooper, John M. 1946a. The Ona. In Handbook of South American Indians, Julian H. Steward, ed. Smithsonian Institution Bureau of American Ethnology Bul. 143, vol. 1:107-25.

———.1946b. The Yahgan. In Handbook of South American Indians, Julian H. Steward, ed. Smithsonian Institution Bureau of American Ethnology Bul. 143, vol. 1:81-106.

Crawford, Meredith P. 1937. The cooperative solving of problems by young chimpanzees. Comp. psychol. Monogr. 14(2):1-88.

———. 1940. The relation between social dominance and the menstrual cycle in female chimpanzees. J. comp. Psychol. 30:482-513.

———. 1942. Dominance and social behavior for chimpanzees in a non-competitive situation. J. comp. Psychol. 33:267-77.

Dole, Gertrude E. N.d. Primal Human Family Structure. ms.

Elkin, A.P. 1954. The Australian Aborigines: How to Understand Them. 3rd ed. Sydney: Angus & Robertson.

Forde, C. Daryll. 1934. Habitat, Economy and Society: A Geographical Introduction to Ethnology. London: Methuen.

Gillman, Joseph. 1939. Some facts concerning the social life of chacma baboons in captivity. J. Mammal. 20:178-81.

Gusinde, Martin. 1955. Pygmies and Pygmoids: Twides of Tropical Africa. Anthropological Quarterly 28:3-61.

Haddow, A.J. 1952. Field and laboratory studies in an African monkey, Cercopithecus ascanius schmidti, Matschie. Proc. zool. Soc. London, 122:297-394.

Harlow, H.F., and Yudin, H.C. 1933. Social behavior of primates: 1. Social facilitation of feeding in the monkey and its relation to attitudes of ascendance and submission. J. comp. Psychol. 16:171-85.

Hallowell, A. Irving. 1956. The structural and functional dimensions of a human existence. Quarterly Review Biol. 31:88-101.

Hebb, D.O., and Thompson, W.R. 1954. The social significance of animal studies. In Handbook of Social Psychology, Gardner Lindsey, ed. Cambridge, Mass.: Addison Wesley, pp. 532-61.

Hooton, Earnest. 1942. Man's Poor Relations. New York: Doubleday, Doran.

Imanishi, Kinji. 1957. Social behavior in Japanese monkeys, Macaca fuscata. Psychologia 1:47-54.

Kempf, Edward J. 1917. The social and sexual behavior of infra-human primates with some comparable facts of human behavior. Psychoanal. Rev. 4:127-54.

Leacock, Eleanor. 1954. The Montagnais "Hunting Territory" and the Fur Trade. Memoir of the American Anthropological Association 78.

———. 1955. Matrilocality in a simple hunting economy (Montagnais-Naskapi). Southwest. J. Anthrop. 11:31-47.

Lothrop, Samuel Kirkland. 1928. The Indians of Tierra Del Fuego. New York: Museum of the American Indian, Heye Foundation.

McCarthy, F.D. 1938-1939 and 1939-1940. "Trade" in aboriginal Australia, and "trade" relationships with Torres Straits, New Guinea and Malaya. Oceania 9:404-38; 10:80-104; 171-95.

Malinowski, Bronislaw. 1931. Culture. Encyclopedia of the Social Sciences.

Man, Edward Horace. 1885. On the aboriginal inhabitants of the Andaman Islands. Reprinted from J. Royal Anthropological Institute (1885). London: Royal Anthropological Institute.

Maslow, A.H. 1936a. The role of dominance in the social and sexual behavior of infra-human primates: I. Observations at Vilas Park Zoo. J. genet. Psychol. 48:261-77.

———. 1936b. The role of dominance in the social and sexual behavior of infra-human primates: III. A theory of sexual behavior of infra-human primates. J. genet. Psychol. 48:310-38.

———. 1940. Dominance-quality and social behavior in infra-human primates. J. soc. Psych. 11:313-24.

Maslow, A.H., and Flanzbaum, Sydney. 1936. The role of dominance in the social and sexual behavior of infra-human primates: II. An experimental determination of the behavior syndrome of dominance. J. genet. Psychol. 48:278-309.

Miller, Gerrit S. 1931. The primate basis of human sexual behavior. Quarterly Review Biol. 6:379-410.

Nissen, Henry W. 1931. A field study of the chimpanzee: Observations of chimpanzee behavior and environment in Western French Guinea. Comp. psychol. Monogr. 8(1):1-122.

———. 1951. Social behavior in primates. In Comparative Psychology, C.P. Stone, ed. 3rd ed. Englewood Cliffs, N.J.: Prentice-Hall, pp. 423-57.

Nissen, H.W., and Crawford, M.P. 1936. A preliminary study of food-sharing behavior in young chimpanzees. J. comp. Psychol. 22:383-419.

Nowlis, Vincent. 1941a. Companionship preference and dominance in the social interaction of young chimpanzees. Comp. psychol. Monogr. 17(1):1-57.

———. 1941b. The relation of degree of hunger to competitive interaction in chimpanzees. J. comp. Psychol. 32:91-115.

———. 1942. Sexual status and degree of hunger in chimpanzee competitive interaction. J. comp. Psychol. 34:185-94.

Putnam, Patrick. 1953. The Pygmies of the Ituri Forest. In A Reader in General Anthropology, Carleton S. Coon, ed. New York: Henry Holt, pp. 322-42.

Radcliffe-Brown, A.R. 1930-1931. The social organization of Australian tribes. Oceania 1:34-63; 206-56; 322-41; 426-56.

———. 1948. The Andaman Islanders. Glencoe, Ill.: Free Press.

Rink, Henry. 1875. Tales and Traditions of the Eskimo. Edinburgh and London: William Blackwood & Sons.

Roth, H. Ling. 1890. The Aborigines of Tasmania. London: Kegan Paul, Trench, Trubner & Co.

Schapera, I. 1930. The Khoisan Peoples of South Africa. London: George Routledge & Sons.

Schebesta, Paul. 1933. Among Congo Pygmies. London: Hutchinson & Co.

———. N.d. Among the Forest Dwarfs of Malaya. London: Hutchinson & Co.

Sharp, Lauriston. 1934-1935. Ritual life and economics of the Yir-Yiront of Cape York Peninsula. Oceania 5:19-42.

Spencer, Sir Baldwin, and Gillen, F.J. 1927. The Arunta. 2 vols. London: Macmillan & Co.

Steward, Julian H. 1938. Basin-Plateau aboriginal socio-political groups. Smithsonian Institution Bureau of American Ethnology Bulletin 120.

———. 1955. Theory of Culture Change. Urbana: University of Illinois Press.

Tylor, E.B. 1888. On a method of investigating the development of institutions applied to laws of marriage and descent. J. Royal Anthrop. Institute 18:245-69.

Vanoverbergh, Morice. 1925. Negritos of Northern Luzon. Anthropos 20:147-99; 399-443.

Warden, C.J., and Galt, William. 1943. A study of cooperation, dominance grooming, and other social factors in monkeys. J. genet. Psychol. 63:213-33.

Warner, W. Lloyd. 1943. A Black Civilization. New York and London: Harper & Brothers.

Westermarck, Edward. 1922. The History of Human Marriage. 3 vols. New York: Allerton.

Weyer, Edward Moffat. 1932. The Eskimos. New Haven: Yale University Press.

White, Leslie A. 1949. The Science of Culture. New York: Farrar, Straus.

Yerkes, Robert M. 1933. Genetic aspects of grooming, a socially important primate behavior pattern. J. soc. Psychol. 4:3-25.

_____. 1940. The social behavior of chimpanzees: Dominance between mates in relation to sexual status. J. comp. Psychol. 30:147-86.

Yerkes, Robert M., and Yerkes, Ada. 1929. The Great Apes. New Haven: Yale University Press.

_____. 1935. Social behavior in infrahuman primates. In A Handbook of Social Psychology, Carl Murchison, ed. Worcester, Mass: Clark University Press, pp. 973-1033.

Zuckerman, Solly. 1932. The Social Life of Monkeys and Apes. New York: Harcourt, Brace.

_____. 1933. Functional Affinities of Man, Monkeys, and Apes. New York: Harcourt, Brace.

Two

EDITOR'S INTRODUCTION

If one of the main stimuli to research in primate social behavior in the 1960s, especially for anthropologists, was the hope that fuller understanding of that behavior might shed new light on the origins of culture, one of the first outcomes of that research was the realization that there is no unitary "primate society" against which a point-by-point evolutionary comparison of human cultural society can be drawn. Groups of nonhuman primates differ greatly in their behavior, across and even within closely related genera and species, and *simple* generalizations about their behavior are rarely to be trusted.

While field primatologists were warning each other against making premature and oversimplified generalizations, behavioral scientists of every disciplinary affiliation were becoming increasingly aware of the work of ethologists such as Konrad Lorenz and Niko Tinbergen. Both of these men argued that behavior patterns, like morphological and physiological attributes, were subject to processes of selection. The ethological approach suggested that observation of a particular item of behavior and comparison of its variations among related species might shed light on behavioral adaptation and evolution. The diversity of nonhuman primate behavior, once it was thoroughly documented and assessed against a background of differing habitats, would provide a key to working out the phylogeny of behavior. Lionel Tiger and Robin Fox argue persuasively in this paper for a comparative approach to the study of "least variable" units of social behavior.

The Zoological
Perspective in
Social Science

BY LIONEL TIGER AND ROBIN FOX[1]

INTRODUCTION

The relevance to a science of new data is not always immediately obvious to
its practitioners. This is particularly so when these data derive primarily from
other sciences and when an admission of their relevance to the science may
require a fundamental critique of its working assumptions. The distinction
between the biological and social sciences has been based on a limited view of
human social action. Though both areas are concerned with social behaviour,
their findings have not been integrated within a comprehensive theory of such
behaviour. A major intellectual system which is capable of achieving this
integration is that based on Darwinian phylogenetic analysis.[2] Over the past
decade a number of studies of behaviour in a variety of disciplines have
recognised the scientific utility of this approach. The work of Washburn
(1962) and his collaborators is well known, and several symposia have been
produced relevant to the general theme of the evolution of behaviour in man
(Spuhler 1959; Howell and Bourlière 1964; Roe and Simpson 1958; Tax
1960). Social anthropologists and sociologists also have become interested in
social origins and behavioural phylogeny (Freeman 1964 and 1965; Geertz
1965; Count 1958; Fletcher 1957; Sahlins 1960). Tentative approaches from
other disciplines have been made towards an understanding of the evolution
of social behaviour, notably by Wynne-Edwards (1962) from zoology,
Ambrose (1965) from psychology, Russell and Russell (1961) from ethology,
and Hockett and Ascher (1964) primarily from an interest in language. A
stimulating attempt at a general synthesis, written for the general reader, can
be found in Ardrey (1961). We do not imply that we agree with all the

Reprinted from *Man* (N.S.), Vol. 1, no. 1, March 1966, pp. 75-81, with permission from
the Royal Anthropological Institute of Great Britain and Ireland.

[1] The overlap between the range of our interests and our names is, we believe, purely
coincidental. We would like to thank Desmond Morris, of the Zoological Society of
London, for his interest and encouragement. Tiger acknowledges the assistance of the
Canada Council in enabling him to take part in this co-operative venture.

[2] "Darwinian" is used here in its widest sense, and is perhaps not the happiest term. We
are aware that on many issues Darwin was wrong, and that there are differences of
opinion among geneticists, zoologists, etc.

findings of all these writers — indeed this would involve the acceptance of outright contradictions — but we cite them to illustrate the range of interest involved which has resulted in the convergence of three strands of scientific endeavour: comparative sociology, physical anthropology, and ethology. Previously social scientists have studied the contemporary or near contemporary behaviour of *Homo sapiens*; physical anthropologists the evolution of human anatomy; and ethologists the evolution of animal behaviour systems. Discoveries in genetics and neuro-physiology provide an understanding of the mechanisms by which not only anatomical structures but also behavioural processes are selected, adapted and transmitted. The convergence of these approaches now makes possible the systematic study of the evolution of human behaviour, and this might require a fundamental change in some basic assumptions about the nature of man and about the nature of social science.

The use of Darwinian insight in social science has been largely analogical. Those social scientists who claimed kinship with Darwin (social Darwinists and social evolutionists) in fact participated mainly in the comparative sociological strand of enquiry. They were not concerned with the evolution of man and behaviour but with the cumulative change of social systems. Contrariwise, the emerging interdisciplinary approach proceeds not from the analysis of social systems *per se*, but from the selection and transmission of genetically programmed behavioural as well as anatomical systems, which are the unitary bases of human social organisation. Thus sociological findings, in this perspective, provide data for a more comprehensive, zoological approach to the evolution of man as a gregarious organism. In consequence the study of human social behaviour becomes a sub-field of the comparative zoology of animal behaviour and is broadly subject to the same kind of analysis and explanation. No special theory other than the Darwinian is necessary to explain the development and persistence of more general features of human social organisation.

Basic to the Darwinian approach is the requirement that the phylogeny of units of behaviour be described, as well as their characteristics. Organisms adapt to their environments and as a result of differential prosperity in survival reproduce selected physical and behavioural characteristics. Human anatomical evolution is increasingly well documented, and because of the connection between physical structure and behavioural function it is possible to describe features of behavioural phylogeny. The key concepts here are survival, adaptation and selection pressure. When faced with any recognisably universal unit of human social behaviour such as dominance/sub-dominance, gregariousness, smiling response, male bonding, greeting, etc., the prime scientific question must be, "what is the function whose selection pressure caused this particular organisation to evolve?" (Lorenz 1966).

Our response to Lorenz's question is facilitated by recent findings in the fields of ethology, palaeo-anthropology and genetics, augmented by work in psychiatry. From ethology we learn of the variety and complexity of animal and particularly primate social behaviour under natural conditions, and may generalise about the relation between organisms, their societies and their environments, in an ecological framework. Here the work of Lorenz and Tinbergen in general ethology needs no elaboration, and, of late, studies of free-ranging primates have improved in number and excellence (de Vore 1965; Southwick 1963). From palaeo-anthropology we derive important evidence about the progressive differentiation of man from other primates particularly with respect to his development of organised predation. The work of Dart and Leakey is central to an understanding of this development, although there are of course many problems in the interpretation of their findings. Genetic research may offer us insight into the mechanism of inter-generational transmission of information not only about physical but also behavioural aspects of the life cycle (Fuller and Thompson 1960; Dobzhansky 1962). Although it was unavoidably unsupported by good biological data, the work of Freud and his various psychoanalytic and psychiatric successors provides us with insights into man's expression and control of his primate nature.

The concern of social scientists is with the explanation of human social action. One way of looking at this is as the outcome of a prolonged process of natural selection, which has produced an animal — Man — with an as yet insufficiently explored repertoire of genetically programmed behavioural predispositions.

DIRECTIONALITY

Lorenz (1958) has said: "The least variable part of a system is always the best to examine first; in the complex interaction of all parts, it appears most frequently as a cause and least frequently as an effect." The burden of our argument is that the least variable part of human social behaviour systems has been neglected because, quite simply, the question of what is "least variable" has been begged. Social scientists have confidently accepted that their knowledge of the left-hand side of the nature/nurture, instinct/learning, heredity/environment, natural/cultural dichotomy is sufficiently firm to permit the relatively unencumbered exploration of the right-hand side. While this was acceptable in an earlier period when suitable data and theory were unavailable, and the indiscriminate use of the concepts "instinct" and "evolution" had been rightly discredited, this question begging is no longer necessary. We are now approaching a position where we can invoke the principle of parsimony along the lines that Lorenz suggests, and to seek to identify those "least variable" parts. Once the *direction* of evolutionary development is known, then the variations can be dealt with. For example, Tiger has argued that such a universal but culturally varied feature of human social behaviour as male aggregation and female exclusion from male groups may in fact derive from a phylogenetically determined need for adult males to form cohesive defensive and foraging units.[3] This genetically transmitted male-bonding behaviour, it is suggested, is of the same biological order for

[3] See L. Tiger, Patterns of male association, forthcoming.

these survival functions as is the male-female bond (however ephemeral) for reproductive purposes. Cultural transmission and social adaptation are clearly responsible for the variety of forms which such aggregations assume; but while these forms are contingent on external pressures, the internal pressure towards their existence in some form is invariant. Furthermore, and this is of particular interest to social scientists, not only their existence, but many of the details of their organisation may depend on genetically programmed sub-systems. Thus the argument that independent sociological or culturo-logical explanations are sufficient to account for even such features of male organisations as group morale, secrecy, female exclusion, hierarchy forma-tion, division of labour, initiation, etc., may require re-examination within the larger perspective we have outlined. This is not to say that conclusive evidence in this direction is yet available — this will ultimately depend on discoveries in molecular biology and neuro-physiology — but that the possibility of such explanation can no longer be set aside. While the argument that social science is not yet ready for such explanations is understandable, the data force us into a position of readiness.

It is well known that Freud foresaw the eventual residence of explanation of behaviour in chemical phenomena. More surprising is Max Weber's comment on the biological basis of *charisma* — a concept central to his sociological thought.

> In the latter field of phenomena [charisma and traditional action] lie the seeds of certain types of psychic "contagion" and it is thus the bearer of many dynamic tendencies of social processes. These types of action are very closely related to phenomena which are understandable either only in biological terms or are subject to interpretation in terms of subjective motives only in fragments and with an almost impercep-tible transition to the biological. But all these facts do not discharge sociology from the obligation, in full awareness of the narrow limits to which it is confined, to accomplish what it alone can do (Weber 1947: 106).

While this may be attributed to a whim or even lapse, it is possible to regard Weber's theory of legitimacy as a complex taxonomic system in the understanding of dominance behaviour. In such a system, "charisma" is essentially a primitive attempt to rearrange dominance orders, or the correlative expression of an intense individual dominance drive of limited incidence in a population (as is, for example, high intelligence). (This does not imply that social stimuli or "releasers" do not evoke and reinforce this drive.) As is known from studies of free-ranging primates, individuals possess differing capacities for the exercise of dominance; consequently there may be profit in reviewing Weber's concepts of dominance, which are, after all, gross and universal categories, in the light of comparably general ethological concepts. In addition, Weber's difficulty in dealing with "illegitimate" forms of dominance may be understood in terms of an analysis of the role of violence in social order along ethological lines.

As an example of the use of both cross-specific comparison and an evolutionary approach, we might cite recent work on the incest taboo. One group of authors (Aberle et al. 1963), in an attempt to answer the age-old question of why man has an intra-familial incest taboo, asked the question, "what other species show similarities to man in terms of, (a) life cycle and reproductive behaviour, and (b) consistent outbreeding?" This led them to look at such factors as asexual imprinting, dispersion activity, intergenerational sexual competition, and the genetics of small inbreeding units. They conclude for man that, ". . . over time, only the familial incest taboo could survive [out of other methods for ensuring outbreeding] because of its superior selective advantages." Another author (Slater 1959), also interested in the possible origins of the taboo, locates it in the demographic features of the "primitive" groups of "early man." The short life-span, small number of offspring reaching maturity, spacing of these offspring and random sex-ratio, made intra-familial breeding a demographic impossibility except on a very small scale. Thus "early man" was an outbreeder in order to breed at all. Later, changes in life-span and family size, etc., made larger scale intra-familial breeding possible, but by this time the "normal" practice had become advantageous in terms of inter-familial co-operation and so was maintained by the taboos (except in some instances where the advantage was not obvious – for example, royal families). These latter points all depend on the dating of "early man" in this argument. We now know that culture emerged long before man had ceased his organic development. As Geertz (1965) puts it, "Tools, hunting, *family organization*, and later art, religion, and a primitive form of 'science,' molded man somatically, and they are therefore necessary not merely to his survival but to his existential realization."[4] (Our italics.) The earlier in this process that a family organisation predicated on outbreeding arose, the more likely is this organisation to have become itself a selection pressure and hence the originator of a possible outbreeding drive. The theory of psychoanalysis, however, claims that an *inbreeding* drive is universal, while Fox (1962) has suggested that intra-familial sexual relationships are basically similar to extra-familial, but that the sexuality of the relationships is subject to extreme variation in terms of differential socialisation experience. The clue to this whole problem perhaps lies in the fact that it is not a simple, specific "outbreeding instinct" which has emerged; it is rather a capacity for self-inhibitory activity – particularly over the sexual and aggressive drives – that has been selected. That the "superior selective advantages" of extra-familial breeding were instrumental as selection pressures in this direction is possible, but the number of variables involved is large and we need to know much more about the "emergence" conditions of early man before we can assess them. As the authors cited earlier note, when discussing the demographic theory, "More primatological and archaeological data may still make it possible to choose more carefully among these assumptions" (Aberle et al. 1963). We put these points to show the kind of leads that may be opened up, when the methods of comparative ethology and an evolutionary

---

[4] See also Geertz (1962).

time-perspective are adopted. The cultural variability of the incest taboo can only be understood when we know what, if anything, is biologically "given."[5]

CULTURAL VARIABILITY

We have tried to emphasise that the "least variable parts" of any social system are variously expressed as a function of the ecological, cultural and demographic situations of the relevant populations. There is no suggestion that learning plays no role or even a minor role in the organisation of behaviour, but rather that there are limits to the plasticity of all animals; that what is *learning* is as important as what is *learned*. As it was graphically put by a colleague — an experimental psychologist — the essential question is, "what's in the rat?" (It is worth noting in this context that rats have found their major survival adaptation in negotiating pipe systems, burrows, sewers, house cavities, etc. That they may be willing and able to master the mazes of experimental psychologists should come as no surprise to the comparative ethologist.)

It follows from this that, within the framework of a zoological theory of social systems, one would expect considerable cultural variability. The result of the interplay of species-specific behaviour with varying external conditions is analysable *once the parameters of such behaviour are known.* It has been demonstrated, for example, in studies of free-ranging primates, that "cultural" differences, i.e., learned and socially transmitted behaviours, exist between groups of the same species adapting to discernably different ecological conditions. In reporting such a study, Washburn and de Vore state that, "the behaviour of the hamadryas baboons shows the function of social behaviour as adaptation, the inter-relations of ecology, troop size, and sexual behaviour, and it suggests the importance of genetics, learning and experience" (de Vore 1965: 612-3).

Primatologists have been able to approach some understanding of the behavioural universals of various species through comparative field studies of colonies under varying conditions. Similarly, an understanding of that syndrome of behaviour which is specific to *Homo sapiens* can only be achieved through the detailed comparative analysis of cultural variability. The data of comparative sociology/anthropology, supported by evidence from the other disciplines mentioned above, are essential, only they would be treated within a different framework of theory than at present employed in these disciplines.

These data are at present collected within a set of non-Darwinian assumptions about the nature of social systems, and research in this vein will have an increasingly limited usefulness for biologically oriented theories. It would be more useful for data to be collected within categories amenable to

[5] In a forthcoming publication Fox will deal with this problem in more detail. (*Kinship and marriage*, Penguin Books.)

analysis by students of nonhuman animal systems. In this way man's social behaviour could be compared directly with that of other species, and interpreted by the same Darwinian concepts. Fruitful areas of research comparable to those developed in comparative ethology might be, for example: territoriality, optimum population maintenance, agonistic behaviour, dominance and hierarchy, bonding, epimeletic behaviour, mating and consort behaviour, ritualised display, play, intergroup relations, communication systems, etc. This expansion of orientation should lead to a better understanding of the non-cultural aspects of human social systems and in consequence to a sharper appreciation of the role of culture in human adaptation. As advances in genetics and physiology are incorporated into this scheme, the distinction between, and the nature of the interdependence of, cultural patterning and genetic programming should become clearer.

CONCLUSION

If we accept that the treatment of comparative zoology is a fruitful one for the analysis of social systems, then we may profitably explore the use of concepts of analysis developed by comparative ethologists. The fact that man is the animal which has relatively recently succeeded in dominating all others does not mean that he is therefore exempt both from being an animal and from being studied as such. Though man's culture is the most evident expression of his biological success over other animals it should not obscure his community with them.

We must stress here, lest we be misunderstood, that we do not advocate an ethological take-over bid for the social sciences. We are less sure of their autonomy than Durkheim was,[6] and as we have seen, even Weber recognised the limitations of a purely sociological approach; but at the same time we must recognise that the basic theories and methods of ethology itself are in a stage of being reconsidered. What we would like to see then is a *marriage* of the two disciplines so that they can profitably explore together the most suitable methods and theories for the study of animal (including human) social behaviour. This article has been written to bring home the importance of ethology to social scientists: a similar one could have been written (and indeed is being contemplated) to bring home to ethologists the fact that, if they are going to study animal social systems, then they have a lot to learn from students of the most complex social animal of them all.

---

[6] We must note, however, that even for Durkheim society was a "natural" phenomenon — a part of nature. Thus, he argued, even though the categories of thought were generated by society they nevertheless "fit" the world of nature outside society. (See Durkheim 1915: 18 and Conclusion.) It is then conceptual thought and language that really differentiates Man from Nature, and concepts are social products; yet society is itself a natural phenomenon.

The working out of this paradox lies behind most of the work of Durkheim's disciple, Claude Lévi-Strauss (1962a, 1962b). The answer surely lies in a proper understanding of the place of thought and language in human social evolution. Concepts are, after all, adaptational devices.

REFERENCES

Aberle, David F.; Bronfenbrenner, Urie; Hess, Eckhard H.; Miller, Daniel R.; Schneider, David M.; and Spuhler, James N. 1963. The incest taboo and the mating pattern of animals. Am. Anthrop. 65:253-65.

Ambrose, Anthony. 1965. The comparative approach to early child development. In Foundations of Child Psychiatry, E. Miller, ed. London: Longmans.

Ardrey, Robert. 1961. African Genesis. London: Collins.

Count, Earl W. 1958. The biological basis of human sociality. Am. Anthrop. 60:1049-85.

DeVore, Irven. 1965. Primate Behavior. New York: Holt Rinehart.

Dobzhansky, T. 1962. Mankind Evolving. New Haven: Yale University Press.

Durkheim, Émile. 1915. The Elementary Forms of the Religious Life. London: Allen & Unwin.

Fletcher, Ronald. 1957. Instinct in Man. London: Allen & Unwin.

Fox, J.R. 1962. Sibling incest. Br. J. Sociol. 13:128-50.

Freeman, J.D. 1964. Human aggression in anthropological perspective. In The Natural History of Aggression, J.D. Carthy and S.J. Ebling, eds. London and New York: Academic Press.

――――. 1965. Anthropology, psychiatry and the doctrine of cultural relativism. Man 65:65-67.

Fuller, John L., and Thompson, William R. 1960. Behavior Genetics. New York: John Wiley & Sons.

Geertz, Clifford. 1962. The growth of culture and the evolution of the mind. In Theories of the Mind, J. Scher, ed. Glencoe, Ill.: Free Press.

――――. 1965. The transition to humanity. In Horizons of Anthropology, S. Tax, ed. London: Allen & Unwin.

Hockett, Charles F., and Ascher, Robert. 1964. The human revolution. Curr. Anthrop. 5:135-68.

Howell, F.C., and Boulière, François, eds. 1964. African Ecology and Human Evolution. London: Methuen.

Lévi-Strauss, Claude. 1962a. Le Totémisme Aujourd'hui. Paris: Presses Universitaires de France.

――――. 1962b. La Pensée Sauvage. Paris: Plon.

Lorenz, K.Z. 1958. The evolution of behavior. Sci. Amer. 199(6):67-78.

――――. 1966. The evolution of ritualisation in the biological and cultural spheres. Proc. R. Soc. (forthcoming).

Roe, Anne, and Simpson, G.G., eds. 1958. Behavior and Evolution. New Haven: Yale University Press.

Russell, Claire, and Russell, W.M.S. 1961. Human Behavior: A New Approach. London: André Deutsch.

Sahlins, M.D. 1960. The origins of society. Sci. Amer. 203(3):76-87.

Slater, Mariam K. 1959. Ecological factors in the origin of incest. Am. Anthrop. 61:1042-59.

Southwick, C.H. 1963. Primate Social Behavior. Princeton, N.J.: Van Nostrand.

Spuhler, J.N. 1959. The Evolution of Man's Capacity for Culture. Detroit: Wayne State University Press.

Tax, Sol, ed. 1960. Evolution After Darwin. Chicago: University of Chicago Press.

Washburn, S.L., ed. 1962. The Social Life of Early Man. London: Methuen & Co.

Weber, Max. 1947. The Theory of Social and Economic Organization. Glencoe, Ill.: Free Press.

Wynne-Edwards, V.C. 1962. Animal Dispersion in Relation to Social Behaviour. Edinburgh and London: Oliver & Boyd.

Three

Vernon Reynolds employs the cross-specific comparative method advocated by Tiger and Fox in a reconstruction of early hominid behavior. Of all the nonhuman primates, the most closely related to man are the African apes. Arguing from the fossil evidence that a forest-dwelling Miocene ancestor common to modern pongids and hominids was behaviorally but not yet morphologically adapted "to bipedal branch-walking and to arm-swinging in the trees," Reynolds suggests that subsequent evolution of the great apes would have involved less behavioral specialization than would that of hominids emerging into a savannah environment. Assuming that "the essential features of the social behavioral patterns of the large apes, which distinguished them from other primates, were probably present in the common ancestor of apes and man, and must have been crucial in determining the modes of adaptation of the emergent hominid line when it left the forest," Reynolds isolates a number of features common to great ape but not to monkey groups and discusses their implications for the evolution of hominid society.

Open Groups in Hominid Evolution

BY VERNON REYNOLDS

In this paper an attempt is made to show that a biological and evolutionary approach to the study of human society can be of value. Humans have species-characteristic behaviour patterns underlying their patterns of social organisation and cultural norms, and these basic patterns have evolved out of the action of environmental selection pressures on the behavioural range of man's ancestral stock. Cultural variability and a range of kinds of social organisation rank among man's outstanding behavioural specialisations and can be seen as the ways in which each society has branched off a basic hominid stem, each society having expanded, modified or adapted aspects of this stem into particular behavioural norms or institutions. In looking at human societies, therefore, we can see a substratum of universal behavioural tendencies manifesting themselves in different forms according to the tradition and ecology of each particular culture.

It is argued here that the typical hunter-gatherer society evolved naturally out of an ape-like system of nomadism, open groups, wide recognition of relationships, sexual differences in temperament and exploratoriness, lack of territoriality, and the inheritance of behaviour patterns such as tool and weapon use, drumming and dancing, and bed-making, which pre-adapted proto-hominids to evolve in certain directions. In addition, it is argued that many of the features typical of present-day societies, such as territorialism, inter-group aggression, rigid structures of authority, strict sexual mores, as well as advancement in creativity and technology, stem from the stage when permanent settlement began and many social instincts had to be controlled and re-directed for the greater selective advantage of the population as a whole.

New finds by palaeontologists concerning the fossil histories of apes and man, together with recent detailed primate field studies, make it possible to take a new look at the probable social and behavioural evolution of man. The biological and evolutionary approach to the study of human society takes the viewpoint that both the individual and the social behaviour characteristic of

Reprinted from *Man* (N.S.) Vol. 1, no. 4, December 1966, pp. 441-52, with permission from the Royal Anthropological Institute of Great Britain and Ireland.

man as a species have evolved as a result of environmental pressures acting selectively on behaviour patterns and range of behavioural variability which are ultimately under genetic control. This does *not* mean that specific items of human cultures, such as marriage ceremonies or tattoo patterns, are inherited direct through the genes: any suggestion on those lines would clearly be unacceptable. It does, however, mean that there is a substratum of inherited behavioural tendencies in man the world over, and that all cultures and systems of social organisation are built on the basis of this substratum. This paper sets out to establish a firm basis for the existence and nature of this substratum, using comparative data drawn from the living *Pongidae*, and justifies this by reference to man's increasingly well understood fossil history.

THE FOSSIL BACKGROUND

The nature of the environmental pressures that must have impinged on the earliest proto-hominids when they began to leave their forest habitat to exploit the savannahs can be readily understood; but we have no direct evidence as to the behavioural inheritance they took with them and which pre-determined them to respond and adapt to those pressures in particular ways. It is possible, however, using data from indirect sources, to build up a probable picture of the kind of creature the forest-dwelling precursor of the first proto-hominids might have been.

In the first place it is necessary to know man's evolutionary position relative to the other primates. Some suggestions based on recent fossil finds by Simons (Simons 1965; Pilbeam and Simons 1965) seem to explain hitherto conflicting and unconnected facts. At the mid-Oligocene site in the Fayum, Egypt, has been found a range of small primates. This appears to include the probable very early ancestors of Old World monkeys, gibbons and the large apes. *Aegyptopithecus*, although it is only the size of a small monkey, is on the line towards the Miocene *Proconsul*, and already has ape-like rather than monkey-like skull characteristics. *Aeolopithecus* may be a very early gibbon ancestor. *Oligopithecus* is fragmentary and of uncertain status but could be related to later Old World monkeys. *Propliopithecus*, hitherto assumed to be on the line towards the gibbons, is thought by Simons and Pilbeam to be too generalised for this, and to be probably related to the larger apes or even the hominid line. The existence of these primate species so long ago, very small and unspecialised, but already differentiated according to ape or monkey taxonomic characteristics, suggests that the ape line and the Old World monkey line may have evolved separately from a progressive prosimian stock in the Eocene epoch, in the same way as did the New World monkeys. The relatively short faces of many of these Oligocene primates, which tend to correlate with trunk erectness, indicates that the prosimian ancestor may have been tree-climbing and erect-postured in the manner of present-day tarsiers.

There is no agreement among taxonomists as to when the hominid precursors diverged from the pongid stock. Some, for example Mayr (1963), basing their opinions on recent comparative examinations of the haemoglobin (Zuckerkandl 1963), the blood proteins (Goodman 1963), and the chromosome structure (Klinger et al. 1963) of the apes and man, conclude that the orang line left in the Oligocene soon after the gibbon stock diverged, and that man and the two African apes have a common, more recent ancestry. Workers on structural morphology, skull topography and details of the hands and feet, such as Schultz (1963) and Biegert (1963), are certain that the hominid line separated from the pongid stock soon after the gibbons, in the late Oligocene or early Miocene, certainly before the three large apes became differentiated. This latter view is the one held here. With regard to the differentiation of the three large apes, it is usually assumed that chimpanzees and gorillas are more closely related than either to the orang. Simons and Pilbeam (1965), however, have recently re-examined all the available dryopithecine fossils and have come forward with some unorthodox conclusions. They are of the opinion that *Dryopithecus (Proconsul) major* is in fact an ancestral gorilla, already distinct as long as twenty-three million years ago; and that *Dryopithecus (Proconsul) nyanzae* was probably ancestral to the chimpanzee. Their latest assessment (Pilbeam pers. comm. 1966) is that in the early Miocene the east African proconsuls spread out across the tropical forest belt of Europe into Asia and that one, *Proconsul nyanzae*, probably gave rise to *Dryopithecus (Sivapithecus) indicus*, which they now regard as a possible fossil ancestral orang, hitherto unrecognised. (It is interesting to note here that Biegert explains the peculiarities of the shape of the skull and dental arch of present-day orangs as being due to an extreme specialised development of the laryngeal sacs, which may be comparatively recent, and not due to an earlier evolutionary divergence.) Thus, the view propounded above is that the ancestral stocks of gorillas and chimpanzees were already distinct in the early Miocene, and that chimpanzees and orang utans have a more recent common ancestry — probably late Miocene — than either has with the gorilla. Behavioural data on the three large apes supports this view. For in terms of social behaviour in the wild, intelligence, learning ability, responsiveness in laboratory experiments, temperament and pattern of juvenile development, orangs and chimpanzees are very similar, whereas gorillas often show marked differences (Reynolds in press; Yerkes and Yerkes 1929).

It would thus seem probable that the divergence of the hominid line had occurred by the Oligocene-Miocene boundary at least. The earliest fossils thought to be on the hominid line belong to the genus *Ramapithecus*, dating from fourteen million years ago in east Africa. The *Ramapithecus* line could have evolved from a proto-ape ancestor towards the end of the Oligocene, prior to the speciation of the proconsul apes.

One criticism of the Pilbeam and Simons hypothesis is that it puts the divergence of the large ape stocks so far back as to antedate the development of those numerous structural specialisations, such as long, brachiating arms, that today seem to be homologous in gorillas, chimpanzees and orangs. The only dryopithecine for which adequate post-cranial matter exists, *D. (Proconsul) africanus*, had arms and legs of equal length, and the build of a monkey. Its arms, however, showed evidence of being relatively free-swinging, and may have been used for hanging, reaching, and leaping in the manner of some New World monkeys. It has been termed a "semi-brachiator" (Napier, 1963). In the evolution of the apes and man it is possible that *behavioural adaptation* preceded the development of structural specialisations in many instances. Thus, according to Simons (pers. comm. 1966), "The common ancestor of the larger apes and man could have been pre-adapted by behaviour, not morphology, to bipedal branch-walking and to arm-swinging in the trees. From this there are two obvious locomotor pathways, one towards increased arm-swinging as in *Pongo,* and to a lesser extent in *Pan* and *Gorilla*, and the other towards human bipedalism. ... The numerous similarities ... may simply be a result of the fact that different stocks, derived from some generalised ancestral hominoid, have had the same basic morpho-system evolved along similar lines because they have the same primary adaptive pattern."

BEHAVIOUR OF MODERN APES

If it is assumed that man and the large apes had a common ancestor towards the end of the Oligocene, and that the hominid line branched off only shortly before the chimpanzee-orang and gorilla stocks diverged, then data on the behaviour of the large apes is relevant to a consideration of man's likely behavioural inheritance. All the large apes have remained within the tropical forest habitat and although all have developed specialisations, they have not undergone the great changes and adaptations of the hominid line which emerged on the savannahs. Thus the essential features of the social behaviour patterns of the large apes, which distinguish them from other primates, were probably present in the common ancestor of apes and man, and must have been crucial in determining the modes of adaptation of the emergent hominid line when it left the forest. The following characteristics are common to the societies of gorillas, chimpanzees and orangs, and are not often found in those of Old World monkeys.

1. *Nomadic.* There is nothing approaching territory ownership in the large apes. None of them has a fixed range beyond which a group rarely wanders and which may be routinely travelled, as in baboons for example. In the case of chimpanzees (V. and F. Reynolds 1965) and orang utans (Davenport in press), movements of individuals and groups appear to be determined by the availability and distribution of food. Large aggregations gather in areas of abundant fruit while in leaner periods solitary individuals or small groups spread out and travel long distances in their foraging. In the case of gorillas (Schaller 1965), which have specialised in shoot and pith eating, food is available all over their terrain throughout the year. But they are also nomadic and travel long distances with little routine in their movements. Although they do return from time to time to the same areas they do not have ranges in the sense that baboons or monkey groups do. No gorilla group has the use of any "core area" exclusively; all are free to come and go as they please.

2. *Open groups and sense of community.* In Old World monkeys an individual belongs to a particular group or is an outsider; in the latter case it would not normally be accepted into a group without fighting. Breeding and all social interactions occur within the group, the members of which are normally always within sight or sound of one another. Although aggregations of groups, such as Hamadryas baboon harems or common baboon troops, occur from time to time in some monkey species, when they disperse again membership of the breeding group or troop remains unchanged (Hall and DeVore 1965). By contrast, all three large apes seem to recognise a far wider nexus of bonds and relationships which could be termed a "sense of community." Chimpanzees and orangs do not live in permanent groups at all. They form temporary associations in which the social bonds appear to be friendship based on like sex or age, sexual attraction, mother-offspring relationships, and possibly sibling relationships (V. and F. Reynolds 1965). One of the most striking features of chimpanzee society is that though the temporary groups split up, the relationships are recognised by affectionate greeting and re-uniting when the individuals meet again. Thus, although the adult and adolescent offspring of a mother leave her to join in other activities with exploring male bands or in sexual groups, they rejoin her from time to time (Goodall 1965 and film). Goodall has observed stranger individuals approach local groups and be accepted in local communities after greeting ceremonies and, in the case of adult males, excited displays. Amongst gorillas, which have become terrestrial, there is far greater stability in group membership. Typical groups contain one or more adult males and a number of females and their offspring. But even in gorillas a sense of community is apparent. For example, some adult males seem to prefer a wandering life, attached to no particular group. Such males are temporarily accepted in established groups without hostility. Sometimes two groups happen to be foraging in the same place and they may join up for a day or two; or they may simply stare at each other and go their separate ways. In either case it is clear that gorillas, like chimpanzees and orangs, recognise ties of relationship which extend beyond the immediate group. It is postulated here that this characteristic of open rather than closed group organisation has typified the main pongid-hominid line since the Eocene prosimian stage, and is responsible for the form taken by human society.

3. *Individual choice in sexual relationships.* In most Old World monkeys the hierarchy of dominance which structures the behaviour of group members controls and limits sexual relations between individuals. In the large apes there appears to be free personal choice. A chimpanzee female in oestrus may solicit and mate with a number of males in quick succession without rivalries ensuing. A gorilla group leader may watch uninterestedly while one of the females of his group mates with a male that has only recently joined them.

4. *Exploratory behaviour of adult males.* In some adult male individuals among chimpanzees and gorillas (and among orangs too, judging from what evidence is available) there seems to be an innate urge to roam and explore. Thus, as was mentioned above, some male gorillas prefer to travel alone most

of the time; on the other hand some seem to prefer the role of "family man" and take on the responsibility of leading a group containing females and juveniles. One can deduce that if new habitats were to become available to gorillas, it would be the roaming males that found them first. A rather similar social pattern is found among chimpanzees. Adult males tend to form small, actively mobile bands of two to five individuals, which travel fast over long distances through the forest and are often to be observed many miles from other chimpanzees. These male bands are real explorers, for they are the first to discover trees newly in fruit, whereupon they call and drum loudly in excitement thus attracting other groups to the area. The exploratory males in this species seem to have evolved a specific function in the social organisation. For in both chimpanzees and gorillas, the females are much less adventurous. Gorilla females are almost never alone, and chimpanzee females, in particular the mothers, are most frequently found in small groups which tend to remain in the same feeding area for days at a time while the adult males are forever moving around. But, as among gorillas, some adult male chimpanzees prefer to remain with the females, and do not seem to move very far. Sometimes these "domestic" males are getting old, but this is not always the case. Among Old World monkey groups there is nothing comparable to the exploratory males found among the large apes. The existence of "bachelor bands" is sometimes reported, but these are usually groups of young males which have retreated from the main part of an organised group as a result of the agressiveness of the dominant males, and their consequent lack of opportunity for sexual liaisons with the females.

5. *Unique behaviour patterns.* Certain remarkable habits are found only in the large apes and man and are evidence of great behavioural plasticity and inventiveness at a very early stage of pongid evolution. Such behaviour patterns include use of tools, use of weapons, drumming and dancing, and the making of beds. For example, gorillas in the wild tear up saplings and plants and hurl them about in an intimidation display; in captivity a gorilla has been reported to use a large and heavy object such as a rock or chair or bedstead as a weapon of attack either by direct hitting or by throwing (Hoyt 1941). Chimpanzees in the wild shake and hurl branches and saplings, and in captivity have been observed to use large sticks as direct weapons against real and dummy leopards (Kortlandt and Kooij 1963). Orangs deliberately throw broken branches down on human intruders in the forest (Schaller 1961; Davenport in press), and an adolescent, home-raised orang spontaneously grabbed a stick and hit at a live snake (Harrisson 1963). With regard to tool-using, both chimpanzees and orangs are known to prepare small sticks to poke into insect holes in order to obtain delicacies; chimpanzees have been observed to use rocks to break open palm nut shells, and have even been reported to fashion drinking cups or sponges from leaves; in zoos and laboratories they become adept at many manipulative skills. Orangs in zoos are notorious for their use of the lever system to prise escape holes (Benchley 1942). Gorillas seem to have almost lost the tool-using propensity, probably as a result of their specialised feeding habits which do not require

manipulative skills, but even gorillas have been taught to paint and draw. Drumming as a form of communication is practised by gorillas (on the chest) and chimpanzees (on the ground and on tree buttresses), and all three large apes occasionally engage in repetitive and rhythmic body movements similar to primitive dancing. Finally, although many birds, insects and mammals build nests, there is no equivalent to the large apes' custom of weaving a nightly bed, except in one or two species of prosimian and, of course, in man, the bedmaker *par excellence*.

IMPLICATIONS FOR THE EVOLUTION OF HOMINID SOCIETY

It is argued that the behaviour patterns listed above, especially the open group system, which are characteristic of present-day large apes and which distinguish their societies from those typical of other non-human primates, were probably present in the common ancestor of apes and man, and that these genetically programmed behaviour patterns determined the social evolution of proto-hominids when they adapted to savannah life. The following is an attempt to reconstruct one logical sequence of hominid social evolution based on the recent fossil chronology and the data on man's ape-like behavioural inheritance.

Out of the proto-ape stock which inhabited the African forests towards the end of the Oligocene a species began to differentiate which specialised in living in the fringes of the forest, where trees are mixed with scrub and grassland. Its members were very adaptable. They were used to bipedal walking and hanging in the trees and were equally at home on the forest floor which they exploited for plants and insects. Curious and exploratory adult males started to increase their amount of animal diet by catching small creatures in the savannah and scrub surrounding the forest, where visibility was better. Sometimes other groups were attracted out of the forest by the excited calling and drumming of the meat-eaters. They had little to fear from predators because they could run fast, quickly climb any nearby trees, and could stand erect to give good visibility over the grass, in addition to being able to intimidate most other animals by their loud calls and hurling of vegetation. Over millions of years this forest-edge species became more distinct in its specialisations and more organised in its social groupings. The male bands began to develop co-operation, and to become more skilled in the use of sticks and stones to kill small animals and to frighten away big cats from their prey. The females and juveniles remained in groups, foraging for fruits and plant foods in the forest and along its fringes. Sometimes such a community of pre-hominids would be widely scattered in and around the forest and out on the savannah. At other times it would be concentrated in areas containing abundant fruit. There were always some adult males, often the older ones, which preferred to remain with the mothers and juveniles in the forest fringes. The roving males normally returned to the trees at night where the community slept in tree nests. Sometimes, when a male band discovered a large source of meat, the fresh carcass of an elephant, for example, they shouted and drummed until other groups, attracted from the forest, joined them. These groups would consist mainly of the younger females and adolescents of both sexes. The noise of the group excitement engendered at these times kept off any potential predators, and some of the groups probably remained on the plains overnight, constructing crude shelters or ground nests of grass and brushwood. Juveniles remained with the mothers on the forest edges and formed play groups of age mates. Out of these playgroups grew bands of adolescents which attached themselves to adult males and made forays onto the plains. But they always returned to their mothers and siblings from time to time. Communities at this stage may have numbered around fifty individuals frequenting particular stretches of forest and plain. Sometimes members would be scattered, sometimes congregated, at other times following nomadic and seasonal routes to known new sources of food. But a community would never be completely separate from neighbouring communities, for bands of males, sometimes with young females, were always travelling from one to another. There were no closed groups in hominid evolution. The transitory stage between proto-ape and proto-man, between forest and savannah living described above, must have occurred over the Miocene period, at the end of which *Ramapithecus* spread out from Africa through Europe and Asia.

Some time between *Ramapithecus* and *Australopithecus* when, during the Pliocene period, the tropical forests were gradually retreating, the proto-hominids became chiefly savannah dwelling. What effects did the ecological pressures of savannah life have on the social behaviour already typical of the new species? Mothers and juveniles were now living on the plains, partially dependent on meat provided by males; with food and water sources more widely spaced, population density decreased. These two facts must have favoured the emergence of more constant groupings than had been the case in the forest and forest edge communities. The most natural grouping to develop was that of a number of friendly or related females and their offspring, accompanied by one or more ageing or domestic-minded males. Gorilla groups may have evolved in the same way when they became terrestrial. These female groups would be scattered around in favourable areas near water holes, where temporary shelters could be made in clumps of trees or bushes, or among rocky outcrops and caves. Their members would spend most of the time foraging for vegetable foods in the vicinity. Together with neighbouring groups they formed a local community from which juvenile and adolescent age mate bands were drawn. In dry seasons communities were large gatherings around remaining water holes. In wet seasons the groups were scattered widely, each finding forage in different areas. Adult males in the communities joined together into mobile roving bands for the purposes of scavenging kills from lions and leopards, finding out new areas of vegetation to exploit, discovering fresh water holes, and making contact with other communities in other areas. These bands might be joined from time to time by young females, and might themselves join up with groups containing sexually attractive females. Males often dragged the carcasses of their finds to a nearby group, or sometimes the groups came out to where the meat was.

Australopithecus and *Homo erectus*. For another few million years the pressures of the savannah habitat developed and structured existing behavioural tendencies. Any increase in the efficiency of inter-individual communication gave advantages to the hunting and scavenging bands, as did any new skill in weapon or cutting tool technology. At this stage it is probable that particular types of stones and sticks and bones were actively sought and retained and primitive fashioning of tools and weapons began. Intelligence was also at a premium as it had always been in primate evolution; for it was advantageous to be able to learn and predict the ways of the other carnivores, the habits of the ungulates, or to remember directions and places previously visited. Development of tools for digging, or receptacles for collecting and of methods of storing plant foods were of survival value; so also were increasingly efficient constructions against wind or rain. The wide ramifications of community ties and relationships and the frequent inter-communications of groups over long distances, encouraged the formation of large co-operative male bands for animal drives and ensured the rapid spread of any new technological development. Already two typically hominid social institutions were clearly in operation: the sexual division of labour, and the basis of the family and tribal systems. The argument here, from our knowledge of ape behaviour, is that the proto-typical hominid family was a matrifocal group of a mother and her offspring, often in association with other friendly or related mothers. The bonds maintaining their cohesion were based on the attraction of the females for each other and each other's young, and not on the common subservience to a dominant male as has often been assumed. While males would act as providers of meat for such a group, the attachment of an individual matrifocal family may have evolved somewhat later. When technology had progressed to the stage where individuals could hunt alone, the smallest economically viable unit at times of maximum dispersion became the nuclear family of a male, a female, and her young. Although males often attached themselves as protectors to these family groups, their role as exclusive sexual partners probably developed much later. The institutionalisation of human tribal systems came at the stage when inter-individual communication in the form of language had evolved to the point where names could be given to designate both the nexus of friends and sex, age, sibling and mother-offspring relationships which were already recognised, and already formed the basis of social interaction over vast areas. It is emphasised that it was the ape-like social organisation of open groups, the network of relationships and the lack of territorial behaviour that caused the evolution of human society with its basic characteristics of extensive kinship systems and inter-group interactions. At no stage did inbreeding, territorial, hominid hordes range the savannahs, being forced to take rational decisions on the subject of co-operation with other hordes — as to whether to marry out or to die out — in order to start human society, as is often assumed. The widespread uniformity of the first stone tool cultures testifies to the truth of this hypothesis. And it is interesting that on quite other grounds than the behavioural ones used here, Vallois (1961: 229) concluded that "All evidence suggests that the Paleolithic bands were not territorial units, that they were capable of large migrations, and that sexual relations must have existed between them."

MODERN MAN AND SETTLED COMMUNITIES

Modern man is territorial and aggressive, hostile to and intolerant of strangers, and lives within an authoritarian social structure in which self-assertiveness and competition for dominance characterises the successful male. If it is true that the essential characteristics of human society evolved naturally out of the adaptation of an ape-like social system to the selection pressures of life on the savannahs, then some additional explanation is needed to account for the advent of inter-group aggression.

As already stated, the evidence indicates that early palaeolithic man was co-operative, not territorial, and had social and sexual relationships over wide areas. Societies still living in a nomadic hunter-gatherer ecology, such as the Bushmen of the Kalahari, or the Hadza of east Africa, show little territoriality or inter-group aggression. Recent studies or re-studies of existing band societies such as the Mbuti pygmies (Turnbull 1966), the American Indians and Australian aborigines (DeVore in press) describe continually changing social groupings, often based on simple friendship or common interest as well as on primary kinship ties. There is little organised leadership at any level; it is, in fact, very similar to an ape-like social organisation. On the other hand, the late palaeolithic hunters of Europe depict hostile bands of warriors and indications of hierarchial tribal authorities in their cave paintings. Probably during the later palaeolithic some populations became less nomadic and made semi-permanent settlements in caves and ravines, becoming dependent upon certain stretches of land containing big game for their survival. In Europe the seasonality of the climate with its long cold winters must have necessitated the sheltering of the community in caves with stored animal or vegetable foods. At about much the same time populations in other parts of the world were starting to develop agriculture, or to follow the migratory herds of ungulates, or to start to tame animals to remain with them.

The fact that the populations which took to permanent settlements became the most successful in the history of human evolution indicates the advantages of settled life, conferred in an increased population growth rate. But the necessity of living close together imposed great strains on the ape-like inheritance of behaviour and temperament adapted to nomadism and fluctuating groups. To draw once more from our knowledge of the behaviour of present-day apes, it is clear that groups of captive apes in zoos and laboratories show social patterns which differ from those found in the wild state (Russell 1966). Social interaction is more frequent and more intense. A hierarchy of dominance is established. Sexual jealousies occur, group structure becomes fixed, individuals may be outcasts and strangers may not be tolerated. On the other hand it is remarkable that only in captive conditions are the real skills, abilities and intelligence of apes demonstrated. Great funds of inventiveness, learning power and ability to acquire new habits are brought out which are never called for in the routine simplicity of their wild lives.

To some extent it is valid to compare the situation of captive apes with that of the successive human populations which took to living in permanent settlements; both are situations of social captivity. As a mode of adaptation such a situation was advantageous in terms of survival and reproduction rate; it fostered the development of unique abilities, too, but it extorted a high price in terms of the social adjustment of the individual with his inherited instincts for quite a different way of life. As a species we have not even yet had sufficient evolutionary time to become adapted to settled living. What were the consequences of settled communities? Chiefly the modifying, institutionalising and rigidifying of existing social behaviour patterns. Thus, a system of permanent hierarchical political authority probably developed from the older men in the family groups now living permanently in the same settlement. In like manner young males of the community were organised into hunting, and later, warrior bands. Juvenile and adolescent age mates from the community formed sub-communities from which emerged hunting bands and female friendship groups. Hunting bands and adolescent groups would still have innate exploratory and social urges, so that neighbouring communities would have continual interaction, sometimes co-operating for particular projects like animal drives, cattle exchange, festivals, or against a common threat. Breeding would be both within and between communities. However, now that communities were attached to particular territories as an ecological necessity, the advent of other tribes, still nomadic, on their land would be viewed with hostility; or, when a community grew too large for its land, a sub-community might break off and search for new land on which to hunt or to cultivate or to graze its domestic animals. In these circumstances territoriality and inter-group aggression began. Thus, most of the social evils of man have probably stemmed from the point at which he became attached to land as an ecological necessity. On the other hand, as energy previously expended on nomadism and constant foraging was saved, so was more effort put into increasingly complex technological achievements.

CULTURAL VARIATION

We have so far followed through some of the possible stages in the evolution of human society out of an ape-like, forest-adapted behavioural inheritance; showing how an open-group social organisation in particular may have pre-adapted and pre-determined the direction of the development of hominid society under the selection pressures of firstly the savannah environment and secondly the start of permanent settlements.

The development of the argument has concentrated on the evolution of those behaviour characteristics which are typical of humans as a species and which are innate, i.e., genetically programmed. As Tiger and Fox (1966: 77) recently stated, "the least variable part of human social behaviour systems has been neglected." In discussing one behaviour pattern — that of the tendency for males to form groups which exclude females — they write: "Cultural transmission and social adaptation are clearly responsible for the variety of forms which such aggregations assume; but while these forms are contingent on external pressures, the internal pressure towards their existence in some form is invariant" (1966: 77). Thus it is argued here that other features found in one form or another throughout human societies — such as political authority (actual if not titular) in the hands of males, attraction of mothers into groups, greater exploration activities of adult males, juvenile and adolescent age mate groups, tribal and kinship systems with recognition of a nexus of roles and relationships, sexual division of labour, and incest regulations (Fox 1962) — are all present in some form or another in all human cultures, and express part of man's genetic behavioural inheritance.

Cultural variations, however, are very real and are also the result of selection pressures. When any indigenous, self-contained culture is studied, it can be demonstrated that the way of life it represents is one possible efficient adaptation to survival within a particular geographical environment. But many factors have prevented even isolated cultures from evolving genetically fixed behaviour patterns specific to the culture. For one thing, measurable cultural variations have been in existence for only hundreds of thousands rather than millions of years. Secondly, for twenty-six or more million years the hominid line has specialised in intelligent, variable response to circumstances rather than a predictable, fixed action pattern. Only this factor has enabled man to colonise new habitats and initiate new behaviour. Thirdly, the typical factors of social organisation, nomadism and exploratoriness (which *are* genetically programmed) have ensured that throughout the evolution of man no social group has ever been totally isolated for long without some inter-change of members, and have thus kept the gene-pool widely homogeneous. Finally, from the stage when permanent settlements and complex cultural variations began to emerge, another factor, social tradition, operated as efficiently as genetic fixing. With the setting up of permanent authority systems and the attainment of true language, skills necessary for the survival of a particular society have been passed on through learning to each new generation. The genetic variability of the group remains unchanged, but if the environmental circumstances alter, or if a more efficient way of doing things is invented, the behaviour of the whole society can change adaptively. This process achieves the same ends as genetic evolution, only much faster.

REFERENCES

Benchley, B.J. 1942. My Friends the Apes. Boston: Little, Brown.
Biegert, J. 1963. The evaluation of characteristics of the skull, hands and feet for primate

taxonomy. In Classification and Human Evolution (Viking Fd Publ. Anthrop. 37), S.L. Washburn, ed. Chicago: Aldine.

Davenport, R.K. In press. The Orang Utan in Sabah.

DeVore, I., ed. In press. Man the Hunter.

Fox, J.R. 1962. Sibling incest. Br. J. Sociol. 13:128-50.

Goodall, J. 1965. Chimpanzees of the Gombe Stream Reserve. In Primate Behaviour: Field Studies of Monkeys and Apes, I. DeVore, ed. New York: Holt, Rinehart & Winston.

_____. Film. Jane and Her Wild Chimpanzees. New York: National Geographical Society.

Goodman, M. 1963. Man's place in the phylogeny of the primates as reflected in serum proteins. In Classification and Human Evolution (Viking Fd Publ. Anthrop. 37), S.L. Washburn, ed. Chicago: Aldine.

Hall, K.L.R., and DeVore, I. 1965. Baboon social behaviour. In Primate Behaviour: Field Studies of Monkeys and Apes, I. DeVore, ed. New York: Holt, Rinehart & Winston.

Harrisson, B. 1963. Education to wild living of young orang utans at Bako National Park, Sarawak. Sarawak Mus. J. 11:220-58.

Hoyt, A.M. 1941. Toto and I. New York: Lippincott.

Klinger, H.P., et al. 1963. The chromosomes of the hominoidea. In Classification and Human Evolution (Viking Fd Publ. Anthrop. 37), S.L. Washburn, ed. Chicago: Aldine.

Kortlandt, A., and Kooij, M. 1963. Protohominid behaviour in primates. Symp. zool. Soc. London 10:61-88.

Mayr, E. 1963. The taxonomic evaluation of fossil hominids. In Classification and Human Evolution (Viking Fd Publ. Anthrop. 37), S.L. Washburn, ed. Chicago: Aldine.

Napier, J. 1963. The locomotor functions of hominids. In Classification and Human Evolution (Viking Fd Publ. Anthrop. 37), S.L. Washburn, ed. Chicago: Aldine.

Pilbeam, D.R., and Simons, E.L. 1965. Some problems of hominid classification. Am. Scient. 53:237-59.

Reynolds, V., and Reynolds, F. 1965. Chimpanzees of the Budongo Forest. In Primate Behaviour, I. DeVore, ed. New York: Holt, Rinehart & Winston.

Reynolds, V. In press. The Apes: Their Scientific and Natural History. New York: Dutton.

Russell, W.M. 1966. Aggression: new light from animals. New Soc. 7:12-14.

Schaller, G.B. 1961. The orang utan in Sarawak. Zoologica, N.Y. 46:73-82.

_____. 1965. The behaviour of the mountain gorilla. In Primate Behaviour, I. DeVore, ed. New York: Holt, Rinehart & Winston.

Schultz, A. 1963. Age changes, sex differences and variability as factors in the classification of primates. In Classification and Human Evolution (Viking Fd Publ. Anthrop. 37), S.L. Washburn, ed. Chicago: Aldine.

Simons, E.L. 1965. New fossil apes from Egypt and the initial differentiation of the Hominoidea. Nature 205:135-39.

Simons, E.L., and Pilbeam, D.R. 1965. Preliminary revision of the Dryopithecinae. Folia Primatol. 4:81-152.

Tiger, L., and Fox, R. 1966. The zoological perspective in social science. Man N.S. 1:75-81.

Turnbull, C. 1966. The Wayward Servants. London: Eyre & Spottiswood.

Vallois, H.V. 1961. The social life of early man: the evidence of skeletons. In Social Life of Early Man (Viking Fd Publ. Anthrop. 31), S.L. Washburn, ed. Chicago: Aldine.

Yerkes, R., and Yerkes, A. 1929. The Great Apes. New Haven: Yale University Press.

Zuckerkandl, E. 1963. Perspectives in molecular anthropology. In Classification and Human Evolution (Viking Fd Publ. Anthrop. 37), S.L. Washburn, ed. Chicago: Aldine.

Four

The Russells' approach to cross-specific behavioral comparison is similar to that taken by Vernon Reynolds in the previous paper; that is, comparisons are drawn on the level of general "societal" patterns and the units compared are seen as adaptive to specific kinds of habitat. However, the Russells, in drawing parallels between the behavior of modern man and that of (primarily) Old World monkeys, cannot stress the biological basis for behavioral similarities in organisms as dissimilar as monkeys and men, hence they phrase their discussion of male and female roles in monkey and human groups in terms of analogous function. The Russells argue that the differentiation of roles on the basis of sex, which may be accentuated among species of monkeys subject to ground predation, aggravates the effect of population stresses, which generate more aggressive and domineering behaviour by the males (both monkey and human).

Primate Male Behaviour and Its Human Analogues

BY CLAIRE RUSSELL AND W. M. S. RUSSELL

Among monkeys living in conditions of relative safety and abundance, males and females tend to be alike in functions, behaviour and status. Among monkeys living in more difficult conditions, males tend to be specialized for leading the band, a function that requires high status, and females for rearing the young; males and females are then brought up in different ways. When monkeys are under population pressure, their social behaviour and male-female relations undergo drastic changes. All these findings have parallels in, and implications for, human behaviour.

Danger and Sexual Differentiation

The ancestors of monkeys probably lived on the floor of tropical forests and in the lower branches of the trees. Evolving from this ancestor, some monkey species have become adapted for life high in the trees. In this safe situation, there is no advantage in fundamental division of labour between male and female, and the sexes tend to behave alike. Other monkey species have become adapted for life in open country outside the forests. Here they are in frequent danger from predatory enemies. Their males, therefore, are specialized for defending the band against predators, for keeping it together, and for leading it along the safest routes to food, while the females concentrate on care of the young. Some monkeys live in the intermediate forest floor situation, and are likewise intermediate in their male-female relations. We will now examine more closely these three general types of monkey societies.

Those monkeys who live in the treetops in forests are exceedingly agile and acrobatic. They can deal easily with predators by concerted mobbing or simply by escape. They do not need to specialize one sex into a defender and protector of the other. Consequently, among these monkeys, there is little or no division of labour. The female takes part in finding food sources and in maintaining the band or family territory against other bands or families; the

Reprinted with permission from *Impact of Science on Society*, Vol. XXI, no. 1, 1971. © Unesco.

male takes part in carrying and rearing the young and in all aspects of their care except suckling. There is little or no status difference between male and female, who are essentially equal in rank. Finally, there is little or no difference between male and female in appearance or physique, apart from the basic differences of sex: they tend to be alike in size and markings.

With some minor variations, this picture generally applies to howlers, night monkeys, titis, marmosets, golden lion tamarins, crested bare-faced tamarins, and gibbons. In all these species the male plays a full part in rearing, though sometimes there is some division of labour within this function. For instance, young golden lion tamarins cling to their mother for the first week or so; they are then taken over by the father and transferred to the mother for feeding. In the fourth month, feeding by the mother stops, but the father continues to carry the young till the end of the month. These forest monkeys show what male-female relations can be like under highly favourable conditions.

The large group of monkey species called macaques includes rhesus monkeys, bonnet monkeys, Barbary apes, pig-tailed monkeys and Japanese monkeys. They live in woodlands, but largely on the ground, and, though they can climb, they are too big, stocky and heavy to be very agile in trees. To some extent, they have specialized their males for functions of leadership and defence against predators. Though the sexes look somewhat alike, the males are bigger, more powerful, and armed with enlarged canine teeth. However, the females are far from being completely dominated, and there are wide variations between societies, even within each species, in the status and functions of male and female. The typical macaque society is illustrated by that of the Japanese monkey.

The complex social structure of a Japanese monkey band is reflected in a complex spatial arrangement when the band is at rest at a feeding-place. This takes the form, literally, of social circles (Fig. 1). The inner social circle, or court, contains young infants (of either sex) and all the females, who enjoy throughout life the special privileges of this court circle — first access to food, and a safe central position when the band moves around. Thanks to this strategic central position, females can exercise an important influence on social organization.

Only a small proportion of adult males live in the inner circle, and then only after spending years on the outskirts. At about the age of 2, young males are relegated to the outer edge of the society as cadets, who serve as scouts on the move. Eventually, a male may be promoted to sub-leader. He then lives on the fringe of the inner circle, and herds the less dominant females on the outskirts of the court, making sure they do not stray or fall behind on the move. Finally, a male may enter the very centre of the court and live there as a leader, controlling the movements of even the most dominant females, emerging to take over from subordinates only in danger situations. The females generally accept being herded by the leaders, but because they can control access to the centre they often determine which males become leaders, and are known to refuse admission to rough, aggressive males. In one

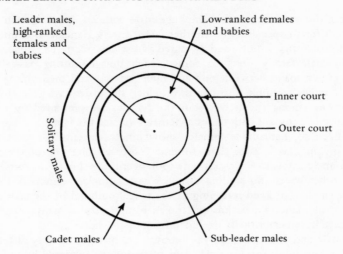

Fig. 1. The social circles of a band of Japanese monkeys, as manifested when the animals are feeding. The basic grouping is that of an inner court and an outer court, with solitary males, social outcasts, living beyond the outer fringes. When on the move, sub-leaders and cadets normally take positions at front and rear of the band.

band, at Minoo, they became dissatisfied with the male top leader, drove him out, and installed the top female, Zuku, as leader of the whole band. She seems to have carried out the normal leadership functions.

Conversely, while Japanese monkey males tend to specialize for public leadership functions, they are certainly capable of parental behaviour, and males who take good care of young have been known to be promoted as a result, because the females let them come to the centre. When the females are about to give birth, their last year's infants are in some cases taken care of by males, who embrace them, carry them around, groom them, and protect them. In a study of eighteen bands in Japan, this procedure was found to be absent in eight bands, occasional in seven bands, and quite regular in three bands. A study of Japanese monkeys in a spacious enclosure in Portland, Oregon, showed that the males become generally more friendly and playful at just this time of year when babies are born. One male, Boris, took charge in succession of two orphan infants and cared for them continuously all year round for at least 18 months.

Cynocephalus baboons live on the ground, usually in open country. There is an imperative need for powerful males for defence against predators. The result is a society ruled by an establishment of leading males, in which females have virtually no influence and are only kept in the centre of the band when they are nursing mothers. Division of labour is sharp, and the male and female look very different.

This physical difference reaches its culmination in hamadryas baboons, where the small female and the huge male with his imposing mane look like two different species. Here another factor enters. Hamadryas baboons live on rocky plateaux where food is scarce and widely scattered. Hence they split into small family units to forage, each unit consisting basically of one full-grown male, between one and nine "wives," and their young. At night, the families assemble into bands, in which the males have generally amicable relations among themselves, but the females are prevented by their mates from social contact with males or females outside the family.

Even so, hamadryas females show innate capacity for leadership. In a colony in Zürich Zoo, Hans Kummer observed that one of the males was extremely inert, so his eldest wife took over leadership of the family, herding the other wives and settling disputes with complete efficiency. To achieve this, she had to keep returning to the neighbourhood of the male and to use his great mane as a backdrop. Human history affords many parallels, where women have been ritually excluded from open leadership.

Male and female hamadryas baboons are brought up very differently, the males remaining for three years in play-groups supervised by young adults, where they learn social know-how, whereas the females are usually kidnapped by their future "husbands" long before this, sometimes before they are 1 year old.

ANALOGIES IN HUMAN SOCIETIES

Now, what analogies with these monkey societies can we find in human societies? Herbert Barry, Margaret Bacon and Irvin Child examined 110 human societies to see to what extent they reared boys and girls differently. They found that the difference in upbringing tends to be large in polygynous societies — those in which one male can have more than one wife — just as in the case of the hamadryas baboons.

It was also found, just as in the case of monkeys, that the extent of the difference in upbringing varies between different societies according to the environmental situation and way of life. Boys and girls tend to be brought up very differently from each other in societies that hunt large game, in those that keep large livestock, and in those that move about a great deal. They tend to be brought up more alike in societies that live largely by fishing in streams or growing root crops. So the differences in upbringing tend to be great when some of the society's tasks demand considerable muscular strength and unencumbered mobility.

The agricultural way of life has been of crucial importance in the evolution of human societies. It is therefore most interesting to see what happens in an agricultural community founded by people who are relatively free from traditional prejudices about male and female roles. This is, in fact, the case of the *kibbutzim* (farm communities) of modern Israel. Those who founded these communities were often highly civilized and enlightened people. They planned to free women from domestic burdens by having communal dining-rooms, laundries and crèches, with men as well as women staffing these services, while women as well as men worked in the fields.

However, most women were found incapable of the heaviest kinds of farm labour, and women when pregnant or nursing were unable to work long hours far from the home buildings. As more and more women had to leave productive farm work, their places were taken by men. Eventually, most women found themselves right back in service jobs.

This strongly suggests that in man's settled agricultural societies there have always been very strong practical pressures, following out of their way of life, to bring up men and women for quite different functions.

In a highly industrialized society, and still more in a highly automated society, all this need not apply. Muscular strength and the capacity for uninterrupted work far from home are becoming less and less generally important. The more advanced human societies are in fact beginning to attain the kind of relaxed and easy conditions that characterize the life of monkeys in the forest treetops — or at least this would be happening if it were not for the pressures of the population crisis.

MALE AND FEMALE SOCIAL PATTERNS

In some forest monkeys, as we have seen, division of labour between the sexes (apart from suckling) is virtually non-existent. When division of labour does occur in monkey bands, it generally takes the form of specializing the male for public, economic, leadership functions, and the female for family, emotional, young-rearing functions. The two sexes often seem to differ in their general pattern of behaviour in a way that accords with their different tasks. Females, accustomed to close and intimate contact with their infants, tend to stay close together in the centre of a band when they can. Males, accustomed to roam about on scouting and foraging missions and to keep out of each other's way in accordance with rank, are generally restless and mobile. They are far more likely than females to leave the band altogether and live alone for long periods.

In connection with their penchant for close contact with others, females are generally more given to grooming their friends' coats than are the males. Among rhesus monkeys (in a zoo), females were found to engage in grooming other monkeys nearly twice as often as males did, among cynocephalus baboons, more than nine times as often, among gorillas, thirty-two times as often as males.

In 1961-63, the islets of LaCueva and Guayacan, off Puerto Rico, were stocked with 278 rhesus monkeys. John G. Vandenbergh studied the way these monkeys formed themselves into bands. He found that "females experienced less difficulty in forming stable social units than males," and that a band usually developed from a "basic group," consisting of three or more adult females and their offspring, which was only later joined by adult males.

The addition of a competent adult male seemed to be essential for the prolonged stability of a band, apparently chiefly because he could put a stop to quarrels. But (according to Vandenbergh) "the basic group of adult females was the most stable portion of the band" and "determined the rank of the band relative to other bands"; this rank remained constant even if one male leader left the band and was replaced by another. The female principle of sticking together thus plays an important part in the creation and maintenance of rhesus monkey society.

In human societies, division of labour also occurs in a systematic direction, and male and female personalities are shaped accordingly. Men tend to monopolize activities that require muscular strength and/or concerted action with other men at some distance from home. Women tend to be allocated tasks that can be carried out at or near home on a routine basis (Table 1).

But, as Table 1 also shows, men also tend to leave to women the service jobs that are not highly regarded, such as repairing clothes, while reserving to themselves the more glamorous manufacturing occupations, even though these require no special strength or mobility.

In accordance with the predominant division of labour, men and women differ, on the average, in their preoccupations and in their permitted fantasy outlets. Men are much concerned with the natural environment, tools and techniques, work relations with other males; women are much concerned with personal, intimate relations. When men feel tense, they are liable to go

TABLE 1.
Distribution of various activities in 224 human societies

Activity	Number of societies in which the activity is always performed	
	by men	by women
Subsistence		
Hunting	166	0
Herding	38	0
Clearing land for farming	73	13
Dairy operations	17	13
Erecting and dismantling shelter	14	22
Tending fowl	21	39
Making and tending fires	18	62
Gathering fuel	22	89
Cooking	5	158
Grinding grain	2	114
Manufacture		
Metalworking	78	0
Boat building	91	1
Making musical instruments	45	1
Work in stone	68	2
Making ceremonial objects	37	1
House building	86	14
Basket making	25	82
Weaving	19	67
Making and repairing clothes	12	95

Source. Simplified from table in a paper by R.G. D'Andrade in E.E. Maccoby (ed.), *The Development of Sex Differences*, London, Tavistock Publications, 1967.

to prostitutes; when women feel tense, they tend to go shopping. In psychoanalytic sessions, men talk of their involvements with the compulsive stimuli presented by women's legs, breasts, bottoms and the clothes accentuating these; women talk and talk about clothes, shopping for clothes, of "having no clothes" and so on.

The differences in interests and fantasies all appear in the dreams, day-dreams, word associations and movie preferences of men and women. These are shown in Table 2, where the male concern with things and techniques and the female concern with intimate relationships are very strikingly revealed.

Eckhard H. Hess and his colleagues have studied changes in pupil size in human beings in response to pictures and have found that the pupils dilate when interest is aroused. They noted considerable sex differences: men responding to pictures of "environmental" interest, such as sharks; women to those of personal interest, such as babies.

By suitable means, they could also plot exactly where each person looked as he or she scanned a picture, as well as noting at which points the subject's pupils dilated. When a picture was used showing a young woman, an older pregnant woman, a man and horse in a ploughed field, with farm buildings and another man-and-horse team in the distance, the men looked chiefly at the bodies — reacting especially to such stimuli as breasts and belly of the women in the picture — and at the distant objects, whereas the women

TABLE 2.

Various observed average differences in interests and fantasies of men and women

Item	Interest or fantasy	
	Males	Females
Subjects appearing most often in dreams of members of 75 tribal societies	Grass Coitus Weapons Animals	Clothes Mother Father Children Home
Most frequent content of most typical day-dream written down by 219 American university students	Money and possessions	Marriage and family
Type of movie most frequently preferred by about 10,000 American children, aged 10-18	Adventure, war and westerns	Romance and tragedy
Kinds of words most frequently produced in word association tests by "a large group of [American] subjects varying in age from early adolescence to old age"	Scientific and business terms Excitement and adventure words Food words	Domestic words Kind and sympathetic words Clothes and colour words

Source. Compiled from E.E. Maccoby (ed.), op. cit.

ignored these environmental features and concentrated on the faces, hence on actual personal relations.

In a recent exhibition of Picasso engravings in London, one of us (C.R.) noted one which showed that the artist had anticipated part of Hess's findings: he had drawn in lines to indicate the direction of gaze of two people: the woman was looking at the man's face, the man at the woman's genitals.

MOTHERS AND SEX-ROLE ORIENTATION

In 1966, one of us (C.R.) drew attention to the predominant behavioural differences between male and female monkeys, in particular to the male tendency to roam and the female tendency to stay together with others. She put forward the hypothesis that these behavioural differences might have their origin in the fact that monkey mothers might react differently to the bodily appearance of their male and female young; in particular they might be likely to repulse their sons earlier and more intensely than they repulsed their daughters, thus initiating the male tendency to roam.

A year later, in 1967, Gordon Jensen, Ruth Bobbitt and Betty Gordon published a study of caged pig-tailed monkeys which provided concrete evidence for the postulated behaviour of monkey mothers and showed that this began very early in the life of the young.

These workers started with the hypothesis that male and female monkeys would behave differently from birth, with the males being more active. They found, however, that in the first few weeks there was no difference in the activity of the sexes. On the other hand, "from the beginning of life," they wrote, "our male and female infant monkeys experienced quite different treatment by their mothers." During the first three or four weeks, mothers held sons closer to them than daughters, and carried them about more.

Jensen and his colleagues suggest that very young male monkeys, like very young male humans, may be less robust than females, and cite evidence that human mothers give more attention to a 3-week-old boy than to a 3-week-old girl.

By the fifth week, however, there was a reversal in the monkey mothers' behaviour. They now carried sons about less than daughters, showed less concern for their sons than their daughters as they moved about in the surroundings, and actively retrieved daughters and held them close more often than they did sons. By this time, too, they were handling their sons, but not daughters, violently, hitting, shoving and throwing them around.

In response to the changing treatment by the mother, the behaviour of sons and daughters also changed. At first, the sons sucked more, but after the fifth week they sucked less than the daughters. For the first six weeks, the daughters played more than the sons did with the inanimate features of the cage. After the sixth week, the sons did this more than the daughters.

In short, males and females behave alike at birth, but males, at an early age, are repulsed from the family circle and driven into the wider social world outside, and this conditions all their subsequent behaviour.

There is evidence that later pressures reinforce this, for the differential behaviour of the mother continues after her offspring are adult. Donald Stone Sade, observing a group of rhesus monkeys, has shown that descendants of a given mother continue throughout life to spend more time and do more grooming with her and with each other than with more distantly related monkeys. However, although sons and daughters both groom their mother in adult life, she apparently grooms her adult sons much less than she does her adult daughters.

Monkey males, too, appear to deal with their very young offspring differently, depending upon sex. In a rhesus colony at the Bristol Zoo, Hilary and Martin Waterhouse observed that adult males played more with male than with female infants. They would even examine an infant's genitals and discard it if it were female. Such paternal treatment could also have an effect in producing sex differences in behaviour.

The behaviour of human mothers towards young children has not been studied as intensively as that of pig-tailed monkeys. However, the results of an extensive study of 379 American mothers are generally in accord with the monkey findings. These women were found to be significantly less affectionate ("warm") towards infant sons than towards infant daughters.

The importance of parental treatment in the very early years can hardly be exaggerated. John L. Hampson studied individuals who had had a change of sex imposed on them at various ages. He found a critical period, coinciding with the development of language – from 18 months to 2 years of age – at which a child's self-identification with one sex or the other became established. Before this age, the rearing of a child could be switched from male to female, or *vice versa*, with little or no resultant psychological disturbance. A change-over made after this period, however, was liable to cause serious disturbances, which were worse the later the change-over occurred. It seems that by about the time he or she begins to talk fluently, a child is "imprinted" with a sense of being male or female, according to the way he or she has been treated by others.

MALE BEHAVIOUR UNDER POPULATION STRESSES

Under relaxed conditions, monkey male leaders attain their high rank by winning friends and influencing people, not by aggressive bullying. In a community of chimpanzees in captivity in very spacious surroundings, it has been found by Vernon Reynolds and Gillian Luscombe that those individuals who rank highest, in terms of access to food, are not the ones who are most often aggressive, but the ones most often involved in friendly contacts. They achieve this by means of a cheerful and noisy display (posturally different from a threat display) which attracts the friendly interest of others, who then approach. In its full form, the display can be done best by males, who thus get the top ranks, for in the wild, with the chimpanzee females burdened with young, the males find fruit trees in season and summon the others. Thus, it is important for leaders to be able to attract attention.

Under crowding, all this changes. In a group of chimpanzees in a small zoo enclosure, the top male achieved his position by brute force and used it to bully his subjects, viciously attacking any who tried to make friendly contact, whether with him or with each other. This reversal, characteristic of all monkeys, can be shown to be a response to crowding, as an indication of population pressure. In monkeys, as in other mammals, the change-over from friendly co-operation to brutal competition culminates in violence, which kills some individuals and weakens others, making them susceptible to disease. In this way, a population in danger of outgrowing its resources is drastically reduced.

Whereas under relaxed conditions, monkey male leaders are chivalrously indulgent and protective to all females and young animals, under crowding stress they become ruthlessly competitive with each other and also with females and young, being quite capable of killing them. In a zoo community of hamadryas baboons, 30 out of 33 female deaths and 5 out of 5 infant deaths were violent killings by males.

Male competitiveness under somewhat less serious population pressure is illustrated by a band of less than 200 Japanese monkeys, which had 6 leaders and 10 sub-leaders. It increased to 440 before it finally split into two separate bands. During the period of increase, however, the number of leaders remained at 6, sub-leaders at 10, with all other males being cadets, the lowest class. There was an attempt by two sub-leaders to break into leader rank, taking advantage of a quarrel between the leaders. But the leaders made up their differences and drove out the rebels, who went off to live as solitaries. The sub-leaders themselves ruthlessly rebuffed the promising cadets who would normally have been promoted, these, too, leaving the band to become solitaries.

Evidently, under population stresses, older males fight to maintain their privileged position and younger males, finding their promotion blocked, suffer from disturbance or despair. In the zoo communities, where escape from the band is impossible, this kind of competition may lead to serious violence.

Under relaxed conditions, then, monkey societies are remarkably peaceful and, so to speak, civilized. Under the stresses of population pressure, these societies become brutally unequal, tense, cruel and violent. The fundamental change is in parental behaviour, from a protective to a competitive attitude. For parental behaviour is the starting-point of all other social behaviour.

Under relaxed conditions, positive parental behaviour develops in two directions. There is a kind of social parental behaviour, directed towards the whole group, which is shown in the constructive and protective behaviour of leaders — for instance, the manner in which chimpanzee leaders search for food and call for others to join them. Then there is a more intimate behaviour, typified by such activities as grooming, which is closely related to the actual care of the young. Often males are specialized in the social kind, and females in the intimate kind of positive parental activity.

It is noteworthy that the former kind of behaviour is more easily reversed into competitive and hostile behaviour under population pressure. Even at moderate levels of crowding stress, males may begin to compete with and attack other males, females and young, though females may still defend their own infants against attack. However, at greater degrees of stress, females, too, reverse their behaviour and may neglect or even kill their infants.

Mankind has rarely enjoyed such relaxed conditions as monkeys can experience in the wild. Technological advances have again and again changed the relation between human societies and their natural resources, and made possible increases of population. But, every time, the population growth has gone on without regulation, so that sooner or later population outstripped resources. Mankind, in the words of H.G. Wells, "spent the great gifts of science as rapidly as it got them in a mere insensate multiplication of the common life." As a result, man has been under virtually continuous population stress, and hence virtually continuous social inequality, tension and violence.

It is easy to find analogues in man of the behaviour we have described in monkeys under crowding stress: competition between males, conflict between the sexes and the generations, domination over and even killing of females and the young. Hence, some people have supposed that all these evils are the normal inherent lot or nature of man.

In fact, in man as in monkeys, social behaviour can be either positive and friendly or negative and competitive, according to the degree of stress. If we make a comparison of human societies under various degrees of stress, we find ample evidence of this principle. In a study of different districts in the city of Newcastle (England), for example, the most crowded third of the city, compared with the least crowded third, produced more than five times as many offences against the person, more than four times as many larcenies, seven times as many people on probation, three times as much juvenile delinquency, more than five times as many cases of neglect of children, and 43 percent more pre-natal deaths.

Man is certainly not more cruel than monkeys, who are capable of great cruelty under stress. On the contrary, man differs from monkeys in having an even more highly developed positive parental urge. This is clear from the long and increasing period of parental care in man and from the unique social achievements of our species, with refinements of social welfare organization even for adults. Even under frightful stresses, where almost all mammals would kill their young, many human beings continue to love and protect them at great sacrifice to themselves.

In man, as in monkeys, positive parental behaviour has developed into social and family activities, with a tendency to restrict the former to males and the latter to females. But if we are to deal with our present problems, we cannot afford too rigid a division between home and work, between family and society, between the old female and male spheres of activity. For we have seen that division of labour according to sex is the product of a difficult environment, and its presence accentuates the effect of other stresses, in particular of population pressure. Now that our technological situation permits us to drop this rigid division and bring up both sexes more flexibly, it is obviously highly desirable to do so.

REFERENCES

Alexander, B.K. 1970. Parental behaviour of adult male Japanese monkeys. Behaviour 36: 270-85.
Hampson, J.L. 1965. Determinants of psychosexual orientation. In Sex and Behaviour, F.A. Beach, ed. London: John Wiley & Sons.
Hess, E.H. 1965. Attitude and pupil size. Sci. Amer. 212(4):46-54.
Jensen, G.D.; Bobbitt, R.A.; and Gordon, B.N. 1968. Sex differences in the development of independence of infant monkeys. Behaviour 30:1-14.
Maccoby, E.E., ed. 1967. The Development of Sex Differences. London: Tavistock Publications.
Napier, J.R., and Napier, P.H. 1967. A Handbook of Living Primates. London: Academic Press.
Reynolds, V., and Luscombe, G. 1969. Chimpanzee rank order and the function of displays. Proceedings of the 2nd International Congress of Primatology, Atlanta, vol. 1, pp. 81-86. Basle: Karger.
Russell, C. Forbidden Fruit. Stockholm: Raben & Sjögren (in press).
Russell, C., and Russell, W.M.S. 1961. Human Behaviour: A New Approach. London: Deutsch.
——. 1968. Violence, Monkeys and Man. London: Macmillan & Co.
Sade, D.S. 1965. Some aspects of parent-offspring and sibling relations in a group of rhesus monkeys, with a discussion of grooming. Am. J. phys. Anthropol. 23:1-18.
Vandenbergh, J.G. 1967. The development of social structure in free-ranging rhesus monkeys. Behaviour 29:179-94.
Virgo, H.B., and Waterhouse, M.J. 1969. The emergence of attention structure amongst rhesus macaques. Man 4:85-93.

Five

The rationale for behavioral comparison along what might be called the traditional ethological axis is that behavior, whether of man or other animals, is to some degree determined by genetic components, hence subject to processes of natural selection and adaptive variation. An important corollary to this assumption is that certain items of social behavior basic to the survival of the individual and, more important, of the species itself are genetically programmed to a high degree, stereotyped within a given species and relatively stable among closely related species. In the case of such species specific fixed-action patterns, cross-specific comparisons of related species of animals can yield (1) insights into the adaptive nature of particular variations and (2) understanding of the behavioral substrate underlying observed variable forms.

In practice, it is far from easy to interpret precisely, in terms of their adaptive significance, differences in behavior between even closely related species of animals as complex in structure and organization as are the anthropoid primates, and when the comparison is of social behavior the problem of elucidating relationships between behavioral and environmental variables is enormous. There is, too, a theoretical difficulty in the way of explaining how natural selection, eliminating individuals from a population, can operate to effectively modify group organizational processes in the face of specific environmental stresses. Not the least of the virtues of the following paper by David Hamburg lies, it seems to me, in Hamburg's observation that "over a long time period natural selection favors behavior patterns that are . . . capable of dealing with highly varied environments" and in his exposition of how the evolution of primate emotional responses, particularly those associated with the mother-offspring relationship, may have tended not only toward prolongation of immaturity but toward the perpetuation of groups organizationally efficient in the protection and nurturance of infants and young.

Evolution of Emotional Responses: Evidence from Recent Research

BY DAVID A. HAMBURG

Why are emotional responses so universal in man and so important in behavior if they have not served some adaptive functions in evolution? Some years ago I presented the view that emotional processes have served motivational purposes in meeting crucial adaptive tasks: finding food and water, avoiding predators, achieving fertile copulation, caring for the young, training the young to cope effectively with the specific requirements of a given environment (1). Now, I wish first to summarize this view, and then to illustrate it with some of the new observations from field and experimental studies of nonhuman primates — especially on the behavior of young monkeys and apes.

Natural selection favors those populations whose members, on the whole, are organized effectively to accomplish crucial adaptive tasks. This is where emotion enters. The sexually aroused mature adult is a convenient example. We can quickly agree that he is quite emotional. In saying that he is emotional, we usually mean that he feels strongly a particular kind of inner experience. In this state, the probability of his achieving fertile copulation is greater than when he is not in this state. From an evolutionary viewpoint, we can further say that he now *wants* to do what the species needs to have done, whether he is aware of it or not. His emotion reflects a state of heightened motivation for a behavior pattern that is critical in species survival.

Viewed in this way, the emotion has several components: a subjective component, an action component, and a physiologic component appropriate to the action. The term emotion usually emphasizes the subjective component — but this is in fact the subjective aspect of a motivational pattern. On the whole, these are motivational patterns that must have had selective advantage over a very long time span. There is probably substantial genetic variability in these motivational patterns, including their subjective components, just as there is genetic variability in every aspect of structure, function and behavior. Natural selection has operated on this variability,

From Jules H. Masserman, ed., *Science and Psychoanalysis*, Vol. XII, pp. 30-52. Reprinted by permission of Grune & Stratton, Publishers.

preserving those motivational-emotional patterns that have been effective in accomplishing the tasks of survival.

It is evident that emotional responses are not limited to behavior that facilitates reproductive success of contemporary human populations. Some nonreproducers may contribute much to the reproductive success of the species as a whole, e.g., physicians and medical scientists who have no children. Moreover, as every clinician knows, contemporary men are quite capable of learning motivational-emotional patterns that are maladaptive by any reasonable standard.

Any mechanism – structure, function or behavior – that is adaptive on the average for populations over long time spans has many exceptions, may respond inadequately to extraordinary environmental circumstances, and may even become largely maladaptive when there are drastic changes in environmental conditions. When we consider the profound changes in human environmental conditions within very recent evolutionary times, it becomes likely that some of the mechanisms which evolved during the millions of years of mammalian, primate and human evolution are now less useful than they once were. Since cultural change has moved much more rapidly than genetic change, the emotional response tendencies that have been built into us through their suitability for a long succession of past environments may be less suitable for the very different present environment.

This point becomes clearer when the time scale of evolution is kept in mind. With present dating methods, the fossil record indicates that mammals have been in existence for over fifty million years, primates for many millions of years, and distinctly human forms for one and a half to two million years. Our own species, homo sapiens, has existed for forty to fifty thousand years. Agriculture came in only eight to ten thousand years ago, and the time is even less – about five thousand years – for effective, widespread use of agriculture. The Industrial Revolution occurred only yesterday in evolutionary time; it is a matter of about two hundred years, less for widespread, effective industrialization. It is only now coming to some parts of the world, and the process of industrialization is rapidly accelerating.

There has recently been an upsurge of research on the evolution of human behavior: new interest, new information and new ideas on a very old subject. There are many sources of information – the fossil record, prehistoric archaeology, chemical dating methods, the few remaining hunting-and-gathering societies ("Stone Age") that still exist, and living nonhuman primates. My emphasis is on broad evolutionary trends, based insofar as possible on regularities in the evidence, converging from various sources. For detailed presentation of relevant research, the reader will find several publications useful (2-10).

Despite the antiquity of the subject matter, its scientific investigation on any substantial scale is very recent. Generalizations are necessarily quite tentative, but the viewpoint is important in trying to understand man.

About a decade ago, Henry Nissen wrote as follows (11).

The behavior of animals is a major contributing factor for their survival and, consequently, through the mechanisms of heredity, for the course of evolution. Maintaining favorable relations with the environment is largely a function of behavior. Possessing efficient skeletal, circulatory, digestive, sense organ, and effector systems is not enough. All these must be used effectively in activities such as food getting, reproduction and defense. Behavioral incompetence leads to extinction as surely as does morphological disproportion or deficiency in any vital organ.

Nissen's statement represents the orientation that has guided my selection of material for this paper.

Animal behavior may be investigated in diverse settings: laboratories, artificial colonies, and natural conditions ("the wild," free-ranging). Information obtained in each setting has its distinctive advantages and limitations. By and large, these approaches are complementary, and all are required for a comprehensive understanding of behavior.

The field studies are extremely helpful in understanding the way that structure and behavior are adapted to environmental conditions; it is really the adaptive aspect that makes the field studies so interesting. For example, the complex relations with other species must initially be studied in the field. The full description of the natural behavior will raise many problems of interpretation that can be settled best by experiments. The richness of primate field studies makes possible a useful interaction between field observations and laboratory experiments; so far there are very few research settings in which it has been possible to pursue a continuous, long-term interplay between laboratory and field studies of primate behavior. We are now in the process of developing such a unit at Stanford.

Primate field studies are very recent. The amount of dependable information in such studies ten years ago was very small; a remarkable acceleration has occurred in the past five years. The workers who have pursued primate field studies have, in the main, come from zoology, anthropology and psychology.

There are about fifty genera and several hundred species of living primates; there are reliable field studies of behavior of only about twenty. The laboratory studies have not used very many species either. The scientific community has been quite limited in utilizing the remarkable variety of primate species. This is one important caution about the present early stage of primate research. For example, the great apes (gibbon, orang-utan, chimpanzee and gorilla) constitute the only family in which most of the genera have been the subject of systematic, long-term observation under natural conditions.

There has been an important and very recent change in what field workers consider necessary to make fruitful observations; at the present time the minimal conditions for an effective study are considered to be about a thousand hours in the field, distributed over a year. Jane Goodall (12), who has done important work with chimpanzees, has so far spent more than four thousand hours distributed over about five years, and indeed much of the critical information came only during the fourth year. In the first year, she saw only little black blobs in the distance at about five hundred yards. Habituation of the animals to the observer is an important consideration; so is the ability of the observer to recognize individual animals and time periods that are long enough to disclose rare but adaptively critical events — these are some of the methodologic problems of the field studies.

Now, with such a variety of species to choose from, what primates are most interesting? This, of course, depends on the question one is asking, and for different purposes different primates will prove to be most suitable. In terms of a general interest in human evolution, it is possible to provide criteria for choice of species upon which to focus. That is to say, certain species are more likely than others to provide interesting leads that stimulate research in the contemporary human species from an evolutionary perspective. The criteria I would accordingly cite are the following: in regard to behavioral criteria, species that spend a substantial portion of their time on the ground; species that show a tendency toward bipedalism (walking upright); species that show a relatively great complexity of learning; and species that show a great complexity of social interaction. Nonbehavioral criteria would include: maximum similarity of central nervous system circuitry to man; maximum similarity in the number and form of chromosomes; maximum similarity of blood proteins and immunologic responses.

Now let us turn to a few general findings emerging from the recent field studies. The newer studies tend to support the concept that, in primate evolution, social organization has functioned as biologic adaptation(1). Both the recent field observations of nonhuman primates and some interesting studies of hunting-and-gathering societies suggest that group living has conferred a powerful selective advantage upon the more highly developed primates. This selective advantage derived from social organization has probably included protection against predation, meeting nutritional requirements, protection against climatic variation, coping with injuries, facilitating reproduction and, perhaps above all, preparing the young to meet the requirements and exploit the opportunities of a given environment, whatever its characteristics may be. I shall return to this point shortly.

All species of monkeys and apes studied so far live in social groups that include both sexes and all ages; there is, however, immense variation in the size and pattern of the groups. The most common groups vary between about ten and fifty, but among gibbons the group consists of a pair of adults and their offspring whereas it may be as large as a few hundred among some baboons and macaques. Group cohesion, inter-animal dominance and patterns of sex behavior are very different in various primate groups, but all monkeys and apes spend the greater part of their lives in close association with other members of the same species(13).

In general, the shift from exclusively tree-living behavior to substantial ground-living activity is associated with the following tendencies: larger home range; larger group size; greater social cohesion; larger body size; lower population density; more aggression — at least in display; more exploratory behavior; more dominance behavior; more sexual dimorphism; increased fighting ability of males; wider geographic distribution of the species; closer relationship of adult male and infant. That is, all of these characteristics tend to increase in proportion to the amount of time spent on the ground. Although there are exceptions, the characteristics listed above appear to be general trends covering many species. The adaptive situation can be summarized as follows: the basic opportunity in the shift toward ground-living is the finding of more food sources and exploitation of a greater variety of ecologic features; the concomitant risk is much greater predator pressure. There are a variety of responses to this adaptive problem; it is a significant area for future field studies.

In the remainder of this paper, let us pursue this evolutionary viewpoint by considering an area in which strong motivational-emotional responses have long been recognized — the interpersonal bonds between mother and infant and, somewhat more generally, the relationships between young organisms and the others who make up their social environment. I wish to look at these emotionally-charged relationships in the light of recent research bearing on their probable adaptive functions in the very long course of human evolution.

In general, among both tree-living and ground-living primates, the infant nonhuman primate is given much special attention and care that ensures its survival (14). Among the various species, there are differences in the relationship of the infant to its mother and to other adults in the group. In ground-living species, the mother and infant tend to stay in the most protected portion of the group and have the protection of adult males, especially when the group is moving across open country; whereas in the tree-living monkeys, such as langurs, the mother and infant may be seen on the edge of the group as often as in its center.

In all species of monkeys for which data are available, it is apparent that the mother-infant relationship is an important part of the groupwide matrix of social relationships. Infant survival depends on the adaptation of the whole group. It is quite clear that the mother with a newborn is the focus of attention; they are very attractive to other group members and the sociometric star is the infant. The infant in most of these species is a distinctly different color in the early part of life, often black; whether this perceptual feature is important in the responses that are elicited from other animals can certainly lend itself to experimental analysis.

The mother-infant relationship is the most intense (in terms of time, energy, and emotion) in the primate group; it outlasts any other social bond. In Jane Goodall's recent work, it is quite clear that a chimpanzee mother's offspring keep coming back to her from time to time up to 10 to 12 years; that is, until they are fully adults, there is important association even though the young have a widening world as they pass beyond infancy. In the chimpanzee situation, which is a very interesting and unusual one, the only highly stable social unit over a long time interval within the chimpanzee community is the mother and her offspring, usually several of her offspring. This is a kind of stable core, a set of stable subgroups within a chimpanzee community(12). Recently, similar observations of enduring mother-offspring relations have been made in a rhesus macaque group on a monkey research island off Puerto Rico.

Let us briefly consider the problem of prolongation of immaturity. Many of the Old World monkeys take several years until they become fully adult. The time required by the great apes is longer, being longest for the chimpanzee: about 10 to 12 years. This maturation period is, of course, longest in humans. Thus, there is a broad trend toward prolongation of immaturity which carries with it certain short-term adaptive disadvantages. There are motor limitations imposed upon the mother by virtue of having to look after her infant; this is quite clear in the baboon and chimpanzee studies. The mother cannot move as freely as she otherwise would. Other members of the group must similarly restrict their activities and make adjustments to take account of the infant's motor limitations. In addition, there are reproductive limitations: not only could the mother move farther and faster if she did not have to carry or walk beside her infant, she could also produce more infants if the period of dependency were shorter. Typically, the adult female primate does not become sexually receptive again until much or all of the lactation period has occurred.

Thus, at first glance it would appear that the trend toward prolonged immaturity in the primates has occurred in the face of some selective disadvantages. What kind of adaptive gain could possibly overcome disadvantages of this sort? The principal gain probably is that this very long protective time can be utilized for learning. It is long enough so that simple elements of learning can be combined in very complex sequences. This learning can be adapted to the specific conditions of a given environment, whatever those conditions may be. *A very diverse array of environments may be adapted to through the shaping of behavior in the long protected interval of immaturity.* This, it seems to me, is the principal adaptive gain that can offset the short-term disadvantages.

If a system of this sort is to work effectively, the mother should be motivated to protect the young; and the young should be motivated to seek her protection. Moreover, the young should be motivated to explore the environment and to learn how to take its salient features into account. It is interesting to ask whether there are indications of an implicit preparation for the environment of adult life. What I am assuming is that, *over a very long time period, natural selection favors behavior patterns that are in fact capable of dealing with highly varied environments.* Let us look at a few aspects of mother-infant relations in this context as a sort of implicit preparation of the young for adult life. Perhaps the most striking observation is the female's interest in and experience with young that will presumably ultimately be useful in the care of her own young. This interest begins very early and goes all the way through the life span in the higher primate species for which relevant data are so far available. It seems as if there is a major sex difference in the attractiveness of the young males and females. While there is also much evidence of male interest in the young, male interest appears more variable than female interest; this variability appears from species to species, and from one time to another in the individual male's life span. Female interest in young is very strong and persistent. For instance, there is much evidence of experience in the handling of very young infants by older female infants, as well as by juvenile and adolescent females. The adult female in the wild is typically an experienced infant handler by the time she has one of her own.

In passing, it is interesting to contrast the adaptive requirements of preparing for a stable environment with those for a rapidly changing one. The contemporary situation, perhaps more than ever before in human history, involves very rapid technologic and social change. Indeed, it is difficult now to predict very much about the future in which our children will live except that it will be quite different from our own circumstances. Under these conditions, the implicit and explicit preparation of the human young seems increasingly to involve preparation for change itself. One important aspect of such preparation for a changing environment is the facilitation of much personal independence. Therefore, it is particularly interesting to examine the data on nonhuman primates for an evolutionary background on the developmental transition from the extreme dependence of the newborn to more independent behavior of the mature animal. In considering such independence, however, it must be borne in mind that it occurs almost always within the context of the enduring, intimate, interdependent primate group, as described above.

In baboon field studies, there is much evidence of exploratory behavior(15). This is manifested in visual searching of the environment and in examination of objects. Over an extended period, such exploration leads to familiarity with the home range; as territory is repeatedly explored, information accumulates regarding food sources, water sources, probable location and behavior of predators, and trees. In experimental studies, maintenance of behavior in nonhuman primates by investigatable rewards has repeatedly been achieved. The following rewards have been effective: manipulative rewards, e.g., a mechanical puzzle; visual rewards; auditory rewards. The basic finding is that primates will work persistently and solve problems in order to see, hear or manipulate (16). This kind of behavior has been studied experimentally by means of a visual exploration box. While the monkey is in this box, he has available a panel on each side, either one of which he can open and look out. Slides or movies may be projected on each side, and the monkey's preference for different types of visual objects can be recorded. For example, the same object can be presented in two different ways in the same time interval, and the monkey can choose which he prefers to watch after sampling each presentation. Different classes of objects have different reward value within each sensory sphere. In regard to visual preferences, bright is chosen over dull, clear over fuzzy, moving over still, and color over black-and-white. Most effective of all rewards in maintaining responsiveness at a high level — i.e., working for the opportunity — are the sights and sounds of other monkeys. In aggregate, these visual preferences seem to have considerable relevance for the natural environment; the monkey prefers the kinds of stimuli that have had adaptive significance for his species in the long course of its evolution. In experiments of this sort, evidence has also accumulated indicating that novelty increases exploratory behavior, whereas fear decreases it. Altogether, from both field and experimental work, there is growing evidence of motivation for exploratory behavior, especially in the young primate.

Such motivation provides a significant impetus for outward movements for the young primate. The field observations suggest a complex regulation of the outward movement of the infant from its mother, which involves integrating the needs for an increase in independent skills with safety from environmental risks, such as falling out of trees, becoming separated from the group, or exposure to predators. This regulation involves the infant and its mother, and also involves other animals, both the males and the females. There seems to be a kind of limit-setting which can be observed, in which many animals participate. The motivation to move outward, to explore the environment, seems to be present and persistent in the infants, though there is much variation in the rate of its development and in patterns of its regulation. In the case of the chimpanzee, the infant's initiative is strongly persistent with respect to weaning, peer relations, relation to adolescents, and examination of objects.

Exploratory tendencies may usefully be linked to observational learning. Riopelle (17) reviewed experiments in which one monkey learns from watching another, e.g., in food choice situations; the observing animal learns from incorrect as well as correct responses of the operator, i.e., from consequences of the operating animal's action. The newer field studies suggest the adaptive significance of observational learning in a social context. Time and again, one encounters the following sequence: (1) close observation of one animal by another; (2) imitation by the observing animal of the behavior of some observed animal; and (3) the later practice of the observed behavior, particularly in the play group, in the case of young animals.

The well-known food-getting adaptation observed by Goodall illustrates such behavior. It is shown very well in the excellent films made by her husband, Hugo van Lawick — both the "termiting" by adults and the observation, imitation and practice of "termiting" by young chimpanzees. The young can be seen sometimes for an hour or so practicing mopping of termites in a clumsy version of the adults' skillful mopping. They also practice picking grass and preparing grass or a twig for termiting. This may occur several hours after they have observed the adults in this behavior. (This termiting behavior is interesting also in the fact that it represents not only tool use but, to a limited extent, toolmaking according to an established tradition.) Similarly, there was much observation-imitation-practice with respect to sexual behavior. Goodall has recorded one young male who directly investigated the genitalia and observed copulation twenty-seven times. The infants mount their peers early and also practice on estrous females who are markedly cooperative in this practice.

In the preceding section, we have been considering the positive, adaptive utility of learning about the problems and opportunities of a particular environment during a protected period of prolonged immaturity. Now let us briefly consider some of the negative, disruptive effects of early interpersonal loss: total social isolation; partial social isolation; permanent maternal deprivation; and brief periods of mother-infant separation.

In the field studies of nonhuman primates under natural conditions, one of the most striking and consistent observations has been the extraordinary richness and diversity of inter-animal contact during the years of growth and development. Recent laboratory investigations with primate species have provided a stark and informative contrast. The most dramatic comparison is provided by the social isolation experiments. (18, 19) The behavioral effects of raising a rhesus macaque until early adolescence in total social isolation — including isolation from contact both with other monkeys and with man — have proven quite devastating. The effects include: gross disruption of inter-animal contact, with emphasis on withdrawal and avoidance of contact; crouching for very long periods with very few responses directed toward the environment; a variety of maladaptive, self-oriented behavior patterns, including persistent thumb-sucking, self clasping and stereotyped rocking; self-punitive behavior, such as self-biting, particularly on approach of other animals. The effects of partial social isolation are similar though less profound and somewhat more reversible. Current research in this area includes an attempt to determine the minimal length of time of rearing in social isolation that will produce these profound effects and a search for conditions under which the disturbance may be reversible.

At the Yerkes Primate Center, Emory University, Davenport and his colleagues are currently studying the effects of rearing chimpanzees in isolation for 2 years (infancy). They find that severely stereotyped behavior patterns are produced by such deprivation. These stereotypes are already present in the deprivation chamber, before the animal emerges after 2 years. They are, however, markedly accentuated by novel and stressful experiences after he emerges. Such animals are being paired with experienced adult females in order to determine whether "therapeutic" effects will occur in response to maternal behavior.

In some of the experiments in Harlow's laboratory, a remarkable degree of compensation for maternal deprivation has been attained by permitting a modest amount of peer play — on a time scale of minutes per day. In the field, the same species plays hours per day, rather than minutes. This contrast adds interest to the experimental finding. The developmental potentialities of peer relations deserve much further investigation.

Several groups, e.g., Hinde's at Cambridge, have been conducting experiments on effects of short-term separation of mother and infant, in the presence of several other monkeys constituting an enduring group. Kaufman and Rosenblum (20) have recently studied the reaction to removal of the mother for 4 weeks in four group-living pigtail monkey infants. All showed distress, manifested for a day by agitated searching and calling, with three going on to a state reported by them as deep depression similar to the anaclitic depression of human infants described by Spitz.

The primate field studies provide some observations of loss and adaptation to loss. In several species primates carry dead infants for days — even a week. It is striking to see a baboon mother carry the dead infant until it is almost shredded; DeVore has recorded such a sequence on film. Goodall has observed a similar sequence in chimpanzees, noting that for a few days the mother often stares at other infants after she has finally given up her own. Infant mortality in primate groups is high; Goodall estimates 50 percent in the chimpanzees. Most of human history probably has been similar in this respect. Thus, there is an *adaptive premium on adequate preparation of those who survive infancy.*

There are many field observations and laboratory experiments with higher primates that suggest the motivational intensity directed toward restoring the mother-infant bond when separation is threatened, and indeed in the face of all sorts of sudden, novel, presumably alarming events in the environment. For example, William Mason (21) has shown in chimpanzee experiments that the greater is the unfamiliarity, the greater is the infant's tendency to cling to mother or mother surrogate. There are many field observations indicating that fear draws mother and infant together with maximal effectiveness. The infant often clings to the mother's hair or to her nipple in a frightening situation. The infant's scream is exceedingly effective in bringing the mother to him; it appears to elicit intense distress in the mother. The early learning of what to fear, what is dangerous in a particular environment, seems to be quite flexible; such fear may attach to different objects and different circumstances in different environments; but these fear commitments tend to be long lasting in the individual life span. This is similar to the experimental literature on the difficulty of extinguishing avoidance responses which have been established under conditions of strong, noxious stimulation. Here, too, fear can be attached to a variety of objects early, but tends to be quite enduring through the life span. (The situation may well be similar in regard to other emotional commitments, e.g., antagonism.) Presumably such early learning of what to fear and how to cope with threatening elements in the environment may be quite adaptive when the environment is stable in its main features. What happens, however, when there are rapid environmental changes within the life span of the individual, as we experience so strikingly in many contemporary cultures? Under these conditions of rapid, periodic environmental change, are persistent fear responses likely to become maladaptive? From a clinical viewpoint, the answer appears to be yes.

There is probably an evolutionary carryover problem here: the species did not evolve under conditions of extremely rapid social changes within the individual's life span. With little precedent, we must now work out ways of meeting such conditions if this species is to survive and flourish.

REFERENCES

1. Hamburg, D. 1963. Emotions in perspective of human evolution. In Expressions of the Emotions in Man, ed. P. Knapp. New York: International Universities Press.
2. Campbell, B. 1966. Human Evolution: An Introduction to Man's Adaptations. Chicago: Aldine Press.
3. Roe, A., and Simpson, G.G. 1958. Behavior and Evolution. New Haven: Yale University Press.
4. Washburn, S., ed. 1964. Classification and Human Evolution. Chicago: Aldine Press.
5. ____ , ed. 1961. Social Life of Early Man. Chicago: Aldine Press.
6. DeVore, I., ed. Primate Behavior: Field Studies of Monkeys and Apes. New York: Holt, Rinehart & Winston.
7. Schrier, A.; Harlow, H.; and Stollnitz, F. 1965. Behavior of Nonhuman Primates: Modern Research Trends. New York: Academic Press.
8. Hinde, R. 1966. Animal Behavior: A Synthesis of Ethology and Comparative Psychology. New York: McGraw-Hill.
9. Altmann, S., ed. 1967. Social Communication Among Primates. Chicago: University of Chicago Press.
10. Jay, P., ed. 1968. Primates: Studies in Adaptation and Variability. New York: Holt, Rinehart & Winston.
11. Nissen, H. 1958. Axes of behavioral comparison. In Behavior and Evolution, eds. A. Roe and S. Simpson. New Haven: Yale University Press.
12. Goodall, J. 1965. Chimpanzees of the Gombe Stream Reserve. In Primate Behavior, ed. I. DeVore. New York: Holt, Rinehart & Winston.
13. Washburn, S.; Jay, P.; and Lancaster, J. 1965. Field studies of old world monkeys and apes. Science 150:1541-47.
14. Jay, P. 1965. Field studies. In Behavior of Nonhuman Primates: Modern Research Trends, eds. A. Schrier, H. Harlow, and F. Stollnitz. New York: Academic Press.
15. Hall, K.R.L., and DeVore, I. Baboon social behavior. In Primate Behavior, ed. I. DeVore. New York: Holt, Rinehart & Winston.
16. Butler, R. 1965. Investigative behavior. In Behavior of Nonhuman Primates: Modern Research Trends, eds. A. Schrier, H. Harlow, and F. Stollnitz. New York: Academic Press.
17. Riopelle, A. 1960. Complex processes. In Principles of Comparative Psychology, eds. R. Waters, D. Rethlingshafer, and W. Caldwell. New York: McGraw-Hill.
18. Mason, W. 1965. Determinants of social behavior in young chimpanzees. In Behavior of Nonhuman Primates: Modern Research Trends, eds. A. Schrier, H. Harlow, and F. Stollnitz. New York: Academic Press.
19. Harlow, H., and Harlow, M. 1965. The affectional systems. In Behavior of Nonhuman Primates: Modern Research Trends, eds. A. Schrier, H. Harlow, and F. Stollnitz. New York: Academic Press.
20. Kaufman, I.C., and Rosenblum, L. The reaction to separation in infant monkeys: anaclitic depression and conservation-withdrawal. Psychosom. Med., in press.
21. Mason, W. 1965. The social development of monkeys and apes. In Primate Behavior, ed. I. DeVore. New York: Holt, Rinehart & Winston.

Six

The papers so far have been concerned at least in part with particular theses relating to human evolution. Sahlins emphasized a specific discontinuity between the behavior of the nonhuman primates and that of cultural man; the others stressed the continuity of evolving primate behavior. John Hurrell Crook, in the paper that follows, is not primarily concerned with drawing parallels between human and nonhuman primate groups nor with extrapolating from data concerning the latter to inferences about the former; rather, he presents an overview of primate studies against a background of current developments in the more general field of ethology. The development which most interests Crook is the "rapidly growing interest in the relations between ecology, population dynamics and social behavior," three areas of research and theory which Crook analyzes in detail as interdependent "perspectives" of an emerging *social* ethology.

Social Organization and the Environment: Aspects of Contemporary Social Ethology

BY JOHN HURRELL CROOK

INTRODUCTION

During the rapid development of ethology in the last decade two diverging lines of research have become particularly apparent. In the first and more voluminous development the classical ethology of Lorenz and Tinbergen has flowered into a rigorous and lively area of study depending fundamentally upon physiological research and the approaches of experimental biology. The main fields of investigation continue to be motivation analysis, developmental studies and the evolution of species specific behaviour. A number of major textbooks presenting this material have appeared. Preeminent among them is Hinde's (1966) remarkable coverage of the subject including an important attempt to synthesize approaches derived from behaviourist psychology with those of ethology.

The second development comprises a rapidly growing interest in the relations between ecology, population dynamics and social behaviour (e.g., Klopfer 1962). The social emphasis here is not so much upon the traditional ethographic study of behaviour patterns shown between conspecific individuals usually studied in dyadic interaction, but upon the relations between individuals and the natural group considered as the social environment within which they live and to which they are adapted. This second development is the subject of the present paper and the field of study will be termed Social Ethology. The links between the socially and the physiologically oriented wings of ethology remain very close, particularly in such areas as the endocrinology of social interaction and developmental studies. These connections illustrate the continuing interdependence of the branches of the subject.

Curiously enough, in spite of an early emphasis by such workers as Espinas (1878), Kropotkin (1914) and Allee (1938), social ethology has not flowered in the clear-cut manner of physiological ethology. Indeed, the recent popular accounts of the subject, avidly read both inside and outside the academic world, far from presenting a stable front of established knowledge, have

Reprinted with permission from *Animal Behaviour*, Vol. 18, 1970, pp. 197-209.

revealed fundamental contrasts in theoretical orientation and interpretation and have given rise to noisy speculation regarding the inferences that may be made from animal to human social behaviour.

One of the major reasons for this lack of clarity lies, I think, in the failure of ethologists generally to consider social behaviour as a group process. Social behaviour has been treated mainly in terms of reciprocal interactions between individuals presenting stereotyped signals to one another, signals moreover that were commonly species specific and which could be broadly considered as innate. Given such material, it is not difficult to interpret such behaviour as the relatively straightforward outcome of neo-Darwinian selection. It would then follow that society, treated as a matrix of such behavioural interactions, is likewise a direct product of natural selection and adapted to the particular circumstances that have moulded it. A number of recent studies of social structure, admirable in other respects, have begun with this a priori supposition. If this were indeed true one could compare societies in the same way as the classical ethologists compared the behaviour of individuals. Unfortunately, societies, being inadequately characterized by such an approach, are unlikely to be programmed entirely in this way as indeed studies of intraspecific variation make very clear.

Although the emphasis on fixed action patterns in communication has been one focus of mainstream ethology an alternative and minority viewpoint has taken a very different stand, one moreover of increasing significance today. The sociologists Emile Waxweiler (1906) and Raphael Petrucci (1906) working in Brussels between 1900 and the First World War developed a well defined social ethology of which sociology referring to man was itself considered a part (Crook 1970a). Petrucci studied social structures as such: spatial dispersion, numbers in groups, group composition, etc., and he pointed out that there were few correlations between a taxonomy of social organizations and the classification of species. At each phyletic level, he saw a marked tendency for similar societies to emerge in parallel adaptation to similar conditions. He concluded that spatial dispersion, group composition and relations between individuals were directly responsive to the environment and that the factors programming the system included such features as food supply, predation and the requirements for sexual reproduction in differing habitats. The limitation on the range of social structures was determined only by the limited variability of the determining conditions. Petrucci concluded that since societies were determined directly by extrinsic factors they cannot be compared in the way that biologists compared morphological characteristics. And it followed, of course, that the social evolution of man was not to be explained purely in Darwinian terms.

Petrucci had taken an extreme position, but the almost total historical neglect of his work is quite unjustified. Recent studies of social organization, particularly in ungulates (Estes 1966) and primates (Crook 1970b), have revealed important intraspecific variations in group composition and interindividual relations which for the most part appear to conform well with contrasts in the ecology of the demes concerned. Such social characteristics, furthermore, are more labile than the patterns of signalling which formerly comprised the main descriptions of social behaviour.

It seems therefore that any statement about contemporary social ethology must begin with at least the following propositions.

(i) Social structure as a group characteristic cannot be conceived as a species specific attribute or property in quite the same way as has commonly been done with, for example, wing colour or leg length. Instead, social structure is a dynamic system expressing the interactions of a number of factors within both the ecological and the social milieux that influence the spatial dispersion and grouping tendencies of populations within a range of lability allowed by the behavioural tolerance of the species.

(ii) Historical change in a social structure consists of several laminated and interacting processes with different rates of operation. Thus while the direct effect of environment may mould a social structure quickly, the indirect effects of this on learned traditions of social interaction come about more slowly and genetic selection within the society even more slowly still.

(iii) Because a major requirement for biological success is for the individual to adapt to the social norms of the group in which it will survive and reproduce it follows that a major source of genetic selection will be social, individuals maladapted to the prevailing group structure being rapidly eliminated. Social selection is thus a major source of biological modification. In advanced mammals it is perhaps of as great an importance as natural selection by the physical environment.

(iv) Lastly a methodological point. It seems desirable in considering group characteristics to shift the research emphasis away from questions concerning informational sources (i.e. relative significance of genetics, traditions, environmental programming, etc.) to direct analysis at the level of the social process itself. This shift would do much to bypass the never ending sterility and unreality of the nature-nurture controversy when applied to social life. An understanding of a social process will in itself help to define the nature of the factors involved in its programming.

From this basis we may attempt a brief conspectus of contemporary social ethology focusing upon three interdependent perspectives.

(a) *Socio-ecology:* the comparative study of social structure in relation to ecology. The main focus here is upon correlations between social organizations and contrasts in ecology.

(b) *Socio-demography:* the relations between social organization and population dynamics including the role of social behaviour as a mortality factor and hence as an important source of genetic selection.

(c) *Social systems research:* the study of the actual behavioural processes that maintain group structure, bring about social change and which may cause the social elimination of some individuals rather than others.

SOCIO-ECOLOGY

Within the last eight years a number of ornithological studies have shown quite clearly that the social structures of species populations correlate closely with ecology. Social structure in fact is one aspect of a whole syndrome of characteristics that are in the broadest sense adaptive. Huxley (1959) used the term "grade" to refer to the characteristic life styles of species adapted to a particular biotype. The species belonging to the same grade commonly show similar social organization. Such correlations have been well demonstrated by research on gulls and terns, penguins, Ploceine weaver birds, Estrildidae, Icterids, Sulidae, and a number of other groups that have been intensively studied comparatively in the field. The work of Tinbergen (1964, 1967), E. Cullen (1957), J.M. Cullen (1960), Stonehouse (1960), Crook (1964, 1965), Immelmann (1962, 1967), Orians (1961), Nelson (e.g., 1967), and Pitelka (1942 and unpublished) comes particularly to mind and Lack (1968) has recently published a major review.

The comparative approach of the ornithologists has now been applied to mammalian groups and has shown especial utility in studies of primates and ungulates. Both these groups are also relatively easy to study in the field, at least when compared with such cryptic beasts as insectivores, bats, rodents and many small carnivores. Furthermore, certain common themes occur in both primates and ungulates suggesting that we may soon have some important integrative principles by the tail.

The primate story due to Altmann, Carpenter, DeVore, Gartlan, Hall, Imanishi, Itani, Kummer, Rowell, Struhsaker, Washburn and other recent workers is perhaps most clearly illustrated by reference to the old-world Cercopithecoidea now extensively studied across a wide range of habitats. Both interspecific and intraspecific population comparisons have been made (reviewed by Crook 1970b).

A survey of current studies suggests a tentative model of socio-ecological relations in these animals. Generally speaking, population demes consisting of one-male groups or "harems," together with peripheral males or all-male groups, occur in a variety of spatial arrangements, in the least ecologically stable savannah and saheal areas having long harsh dry seasons and relatively low predation frequency. In ecologically more stable woodland savannah and light forest with less extreme annual climatic fluctuation and presumed to have higher potential predation frequency multi-male troops are common. Finally, in tropical and semi-tropical forests an increasing number of species have been found once more to live in small one-male groups often confined to small territories. Peripheral non-social males also occur and recent studies suggest a high rate of male interchange between groups, the female membership of which remains relatively constant.

It has been argued that this range of social contrasts is adapted to differences in the food resources and their seasonal availability in the differing habitats. In the more arid areas food resources are limited in the long dry season at which period dispersion in small groups, the reduction of males to a singleton in reproductive groups and the separate foraging of all-male groups allows optimum food availability to females which are commonly pregnant or lactating. The most effective evidence, although not conclusive, comes at present from the Gelada baboon in Ethiopia. These data are furthermore unconfounded by grouping tendencies influenced by the shortages of safe sleeping sites (Crook 1966; Crook and Aldrich-Blake 1968).

Geladas live near canyons and sleep on the gorge cliffs. During the day they wander either in harems or in congregations of independent harems and all-male groups that may together number several hundred. In the rains the animals move slowly in large herds over rich food resources. As the dry season progresses and food shortage becomes visibly apparent, Geladas travel rapidly in small groups often of single harem or all-male group size and the greater part of each day (e.g., up to 70 per cent of all afternoon activity) is spent in feeding. At this season all-male groups tend to disperse away from the canyon more markedly than do the harems. This appears to reduce the food competition between the two types of group. As we shall see such behaviour also occurs in certain ungulates and leads to a differential mortality between the sexes, females being far more fortunate than the non-reproductive males.

In richer savannah the large troops of baboons or macaques live in areas where food resources normally appear to be sufficient to withstand sustained exploitation by sizeable groups and furthermore such groups offer increased protection against the many potential predators.

The discovery of one-male groups in forests poses difficult theoretical problems but we may perhaps employ in this primate context Ashmole's (1961, 1963) argument applying his explanation of sea-bird breeding biology to the reduced clutch sizes of forest bird species when compared with those of open country relatives (see Lack 1966, pp. 266-270). From Ashmole's viewpoint we may argue that the relatively unchanging forest environment, together with the high density of individuals present due to the great productivity of the biotope, produces conditions in which the population is very close to food shortage throughout the year. Under such circumstances it would be advantageous for individuals to live in small spatially circumscribed groups with low male representation. This would once more permit an optimum allocation of resources to reproductive females.

As Aldrich-Blake (1970) has remarked this view is certainly a simplification. While productivity in tropical forests may vary relatively little, this is not true within the home ranges of individual groups. In any one range there are relatively few food resources, say fruit trees, and these are in fruit only periodically. Feeding conditions, even given a wide variety of plant foods, are likely to vary greatly from week to week. Nevertheless he considers that feeding ecology seems very likely to be involved in an explanation of the contrasting spatial dispersions of forest monkey species. Current studies of the autecology of certain species and the synecology of the monkey populations of given forests will go far to explain these differences.

It remains plausible, however, that social forces alone may play a greater role in the determination of primate social behaviour than is at present known. The Gelada and the Hamadryas baboons (Kummer 1968) live under quite comparable conditions and have social structures that at first sight resemble one another closely. Yet the fine details of group dynamics and the individual qualities that make for reproductive success differ greatly. Adaptation to contrasting group processes within groups of comparable structure may lead to the selection of animals of very different behavioural character. Contrasts in social processes within similar social structures may thus be based on differences in behaviour that are the results of marked and long-term social selection.

Recent work on a wide range of African ungulates by Jarman (1968) reveals a fascinating range of social structures that partially parallels those of primates. In forest and forest fringes most species such as duikers and dik-dik are small in size, show little sex dimorphism, show no sexual segregation, have territories and live in pairs or alone. These animals are browsers. Other forest and savannah forms, oribi and bushbuck, for example, and also a number of montane animals not treated by Jarman live in small groups of up to about twelve. Males are a little larger than females. Some show territorialism and all are grazer-browsers. In savannah and open savannah the social units are various, usually consisting of one-male breeding groups, all-male groups and sometimes lone males. The one-male breeding groups may be territorial or move as parts of herds. In some species such as the wildebeeste (Estes 1966) some populations are migratory and gregarious, others nearby in a more stable ecology being stationary and territorial. These animals are browser-grazers and grazer-browsers. Out on open grassland plains occur vast herds of mixed sexes and ages (eland, buffalo) in which several varieties of one-male group or multimale reproductive units occur. These are mostly of nomadic grazers.

Jarman interprets the survival value of these various grouping tendencies mainly in terms of food availability, particularly in relation to seasonal differences in resources. For example, he argues that the dispersed condition of many forest browsing ungulates is related to the spatial and temporal scattering of suitable food plants. The large nomadic herds of grass plains by contrast are a function of the need for protection from predators in an environment offering little cover and which allows large congregations to form. Nomadism is related to the danger of over-exploitation of resources in a given locality, the large expanse of available habitat and the seasonality of the richness of the grass cover.

Jarman's (1968) study of species with one-male reproductive units living along the Kariba river is especially important for he shows that in the dry season the one-male groups occupy the food-rich riverine areas while the excess males are dispersed further inland where they get less food and are subject to higher predation. He demonstrates that such a dispersal also means that a predator has a higher probability of encountering a male than a female prey animal. Clearly the one-male group structure is highly advantageous to the individual females and single males living in the breeding unit and is a direct cause of much non-breeding male mortality.

We cannot yet tell whether all mating systems of one-male group living ungulates necessarily have these effects. Jarman has nevertheless contributed important evidence for the involvement of a social factor in population dynamics, which goes a long way to vindicate arguments used in explaining the similar social structures in primates.

SOCIO-DEMOGRAPHY

The dispute in socio-demography has centred upon two alternative viewpoints. Lack (1954, 1966) originally interpreted population dynamics in birds as largely the outcome of interacting density-dependent environmental factors. He argues that spatial dispersion is a consequence of the natural selection of individuals and that it allows maximum recruitment from breeding units. By contrast Wynne-Edwards (1962) has argued that social behaviour influences dispersion to maintain optimum numbers in relation to resources, that dispersion patterns are a consequence of group selection and that socially mediated mortality is the prime factor in population dynamics.

In the hands of the main protagonists these approaches have perhaps both tended to acquire somewhat scholastic attributes. There is in fact a serious dearth of critical studies and Chitty (1967) has called for a more open-ended theoretical and a more experimental approach to the whole problem.

Wynne-Edwards' work has focussed attention upon the question whether social attributes promoting dispersion do in fact play a major role in population dynamics as mortality factors and therefore also as important social selection pressures in evolution. A demonstration that this may often be the case would not however necessarily support Wynne-Edwards' theory as a whole. Indeed it would allow (see below) a construction of "open adaptive" models (Buckley 1967) of the socio-demographic process, perhaps representing quite closely Lack's more recent position (1968).

TERRITORY AS A SOCIAL MORTALITY FACTOR

Some six studies on a variety of birds appear to demonstrate the significance of territory in this respect. Jenkins, Watson and Miller (1963) working on grouse and ptarmigan, Delius (1965) on skylarks, Tompa (1964) on song sparrows and also Patterson (1965) and Coulson (1968) on certain colonial seabirds have all published material strongly supporting Kluyver and Tinbergen's (1953) original views concerning the great tit.

The studies of grouse, song sparrows, skylarks and Dutch tits all indicate that either the autumnal or spring occupation of territories causes a dispersion of the population in excess of the carrying capacity of the local habitat. The excluded birds may either leave the location and suffer increased mortality through various causes in the peripheral and sub-optimal environments or they may remain as a non-breeding population overlapping with the breeding population and to some extent perhaps competing for food and ready to occupy vacated territory as soon as it appears.

The work on gulls has shown that there is survival value in terms of reproductive success in occupying the preferred territories in the centre of a colony rather than peripheral sites. Coulson in particular showed that intense competition for central sites in a kittiwake colony resulted in their occupation by heavyweight birds. Both sexes surviving at least 5 years from the time of first breeding were heavier there than were neighbours that died within this time. In the centre of the colony there was also a larger mean clutch size, higher hatching success and more fledged young than at the periphery.

It appears that spatial dispersion whether in conventionally territorial or colonial-territorial mating systems involves higher individual survival and better reproductive success for the occupiers of prime sites than for those animals forced to a periphery.

However, while the breeding stock of these species was commonly determined by dispersal, this was not always invariably the case. For example, on food-poor grouse moors, numbers varied more annually than on food-rich ones, suggesting a more direct control there by extrinsic environmental factors involving food shortage. Similarly with the skylarks, in one spring following a harsh winter the population was well below the carrying capacity of the local preferred habitat and no dispersal effects were operative that year.*

These findings recall Kluyver and Tinbergen's (1953) report that the numbers of great tits vary annually more in less preferred woods than in the most preferred areas nearby. Working at Oxford Krebs (personal communication) reports that there is now evidence that the territorial factor may be more significant in the control of tit numbers than had been earlier supposed.

SOCIAL MORTALITY IN FLOCKS

Some years ago Lockie (1956), working then with Corvids, argued that individual distance and dominance-subordination phenomena in bird flocks had survival value in that in an encounter over a food item the loser could avoid its opponent without damage and the winner would win without a fight. Either way energy valuable to the individual is conserved. Under conditions of food shortage, however, the loser will progressively starve. The accumulative loss to a population mounts gradually, the relatively subordinate dying first. Were all birds equal presumably the cutback would be sharp and sudden with little mortality differential.

The work of Murton, Isaacson and Westwood (1964) has shown very clearly the importance of such socially mediated mortality in British woodpigeon populations. In flocks the feeding rates of these birds are greater in the middle and to the rear than in the van. Only a proportion of the pigeons can gain entry to the preferred flock centre. Others are pushed to the periphery. Those in front, harried by those behind them, eat less and commonly flee from flock to flock, usually again landing in front and being hustled. Under limiting conditions the effect of such behaviour leads to differential mortality.

Murton et al. (1964) note that the social effects allocating individuals to starving or non-starving sections of a population do in fact adjust flock size within limits to food availability and do maintain the highest survival rate

---

*Conceivably this may always be the case in more northerly populations, subject to hard winters, annually.

relative to supplies. However, the system sometimes breaks down. One year persistent heavy snowfall forced the pigeons into sudden conflict for limited Brassica plants. Then almost the whole population suffered a severe weight reduction at the same time. The effectiveness of social factors in limiting mortality thus depends on food type, item size and dispersion.

Comparable findings emerge from studies of the African savannah weaver bird, *Quelea quelea*, studied in the field by Ward (1965) and in the laboratory by Crook and Butterfield (1970). Ward, an ecologist, had shown that during the dry season period of reduced food availability these weavers show a rapid drop in weight, particularly marked in females. At the same time the proportion of males in the population increased. Ward concluded that competition for food led to an elimination of females. Laboratory work shows that males are dominant over females in mixed sex groups and experimental studies suggest that oestrogen inhibition of the LH effects otherwise maintaining a low threshold for aggression in males is responsible for this sex difference. An effect of male dominance appears to be a major reduction in the number of females ready for reproduction at the onset of the breeding season. Unlike many species of weaver, the *Quelea* is obligatorily monogamous yet, even so, many nests in colonies are never occupied. It seems that in the rather harsh Sahael environment where *Quelea* lives the practice of monogamy allows a male to assist the female more in rearing the brood than would be the case had he several mates. Both his and the female's reproductive success may indeed be maximized by the procedure. The natural selection of the behavioural features determining this monogamy are likely to have arisen within the context of the shortage of available females for breeding. And this, as we have seen, is a consequence of differential socially-mediated mortality between the sexes in the dry season.

In general, the available studies show that the control of numbers is brought about by numerous factors, some environmental and some social. At any one period the key factor involved may be extrinsic, for example, food shortage, while under other conditions social factors such as territorial behaviour may be the prime regulator through their effects on dispersal. Sometimes, one may suppose, that several factors interact to produce a given outcome and that no clear-cut key factor is operating. In addition the effects of an extrinsic factor may be mediated or buffered by a social process such as the intra-flock spacing mechanism to produce gradual and selective mortality rather than an abrupt fall.

Such a process of population control may be relatively easily modelled using a computer and allocating arithmetic values to hypothetical "Availability to Demand ratios" for each of the several commodities likely to act as controlling factors. Such a model is open and adaptive. A relative steady state is maintained by virtue of the limited variance of each factor. However, should one or more factor move by stepwise change to another range of variance, such a population may be conceived as adjusting to the new levels or new ecological "legislation," as Solomon (1964) calls it.

Such a control model, inherent in the views, for example, of Chitty and his associates differs markedly from the closed homeostatic model used by Wynne-Edwards. In the latter, the animal's perpetual calculating of the relation between its own numbers and its resources, even including those it has yet to exploit, functions as a Sollwert giving the fixed point about which the homeostat functions. There is in fact little evidence for such a model and the one proposed here appears to concord best with reality.

THE SOCIAL PROCESS

The traditional approach of ethology to social interaction has consisted in the study of reciprocal behaviour usually between members of dyads and the signal patterns used in such behaviour. The dyadic relations are classified in terms of context, courtship, mating, etc., and the sum of such features (an ethogram) often treated as an adequate account of the social process. At least with advanced mammals, this now appears most unlikely to be the case.

Imagine attempting to understand a game of football by means of a study of the dyadic interactions between individual players. As Ray Birdwhistell has repeatedly emphasized, such an approach is sterile. To understand the game, the social location of each individual as a role player in relation to each and every other player needs description. Then, with the ball in motion, the relations between these relationships become apparent and the rules may be determined. To gain comparable information for behaviour within mammal societies is an exacting task but one in which considerable progress has been made recently, using both wild and captive groups of primates, mostly macaques. To conclude this paper a brief account of the current perspectives in social primatology will be presented, drawing mainly on the recent work of Japanese, American and European workers.

For many years the structure of primate groups was analysed primarily in terms of dominance. The existence of a status hierarchy was generally thought to stabilize relationships through the reduction of social tension, each animal knowing its place. Often animals are found to cooperate either in the enforcement of existing rank relations or, by contrast, in upsetting them. Cooperation in social control emerges as an important problem area in primate social research. Recently it has become apparent that the simple dominance concept was not only inadequately defined and carrying many unwarranted motivational overtones but that the description of group structure in dominance terminology was in many species not only a difficult task but also an inappropriate procedure (Gartlan 1968).

In an important discussion of dominance in a captive baboon group Rowell (1966) infers that relative rank depends upon a continuous learning process in relation to rewards in inter-individual competition for environmental or social goals. This occurs, moreover, against a background of differential kinship status and the observational learning of the behavioural styles of companions. Relative rank is much affected by health. Dominance ranking is based upon the approach-retreat ratio in encounters between two individuals in a group. Measures of rank by differing criteria do not, however, necessarily correlate and Rowell found no single criterion for high rank.

Rowell (1966) shows that apparent rank is a function of the behaviour of the relatively subordinate. Higher rankers, at least in Cercopithecoid primates, evidently feel free to initiate interactions. These initiations commonly lead to some suppression in an ongoing activity by a subordinate or to an outright conflict. Subordinates learn to avoid such situations. Avoidance learning leads to behavioural restraint that leaves higher rankers even greater freedom of movement, easy access to commodities and freedom to initiate behaviour with others. In competition for commodities in short supply, low rankers are likely to suffer deprivation and in social relations repeated constraint may involve physiological "stress" and concomitant behavioural abnormality.

Hall and DeVore (1965) describe the "dynamics of threat behaviour" in wild baboon troops. A male's dominance status relative to others is a function not only of his fighting ability but also his ability to enlist the support of other males. In one group studied two adult males formed a central hierarchy, the pivot around which the social behaviour of the group was organized. When one of these two died the remaining one was unable to prevent the third ranking male in cooperation with a newcomer (a subordinate male that had left another troop) from establishing themselves as central. The third male and the newcomer had evidently become affiliated when both had been relatively peripheral in the group structure. Common mutually supportive behaviour seems to have been the precondition for their "success" in assuming high rank later. By so doing they gained the freedom to express behaviour in the absence of previous constraints and to initiate behaviour as and when they wished.

Wilson (1968) provides further information on mutual support in a study of the rhesus troops on Cayo Santiago Island. Young males tend to leave the smaller groups and move into the all-male peripheral areas of larger ones. When they do this they are commonly attacked unless they gain the protection of another male already established there. It so happens that males that give support are usually relatives, even brothers, who originated from the same natal group as the "protégé."

The inadequacies of the dominance terminology have led Bernstein and Sharpe (1966), Rowell (1966) and Gartlan (1968) to describe the social positions of individuals in a group in terms of roles. Roles are defined in terms of the relative frequencies (e.g., per cent of group occurrence) with which individuals perform certain behavioural sequences. When the behaviour set of an individual or class of individual is distinct the animal is said to show a "role."

Bernstein (1966) emphasizes particularly the importance of the role of "control animal" in primate groups and shows that such a role may occur in a group of capuchins, for example, in which no clear status hierarchies can be established. The prime responses of a control animal are assuming a position between the group and a source of external disturbance or danger, attacking, and thereby stopping, the behaviour of a group member that is distressing another, and generally approaching and terminating cases of intragroup disturbance. Whether or not a control animal is also recognizably the "dominant" or a "leader" (in the sense of determining direction of march) depends upon the social structure in which he or she is situated.

Social position in primate groups may be well described in terms of roles but little attention has been given as yet to an appropriate set of descriptive terms. It is one thing to say an animal shows a "role," another to say precisely what is meant. Using concepts derived from writers such as Nadel (1957) and Sarbin (1959) we may describe a primate's social behaviour in terms of the individual's age and sex status, social position and group type affiliation. In any given group each individual shows characteristic patterns of response in relation to others in the group, to older animals, to dominant animals, to subordinates, to peers of comparable kinship, rank, etc. The range of an animal's behaviour patterns in relation to others in the group comprise its social behaviour repertoire, as shown with its companions in the group. The sum of the proportions of group scores of such characteristics shown by an individual defines its social position in the group. An observer may prepare a social position matrix by allocating the relative frequency of interaction patterns shown by each member to a particular cell. In a more general statement each individual may be defined by its proportion of the total of the various interaction patterns shown within the group.

It so happens that macaque social positions may be categorized into consistent types that recur repeatedly in new groups formed from the division of the old or in separately analysed independent groups. Such categories are termed "roles." There is, for example, the control male role, the central sub-group secondary male role (competitive with another in the sub-group, dominant to peripheral males but subordinate to the central animal and commonly supportive of him), the peripheral male role, the isolate male, the central and peripheral female roles. We may also bracket together certain types of behaviour to describe roles of animals of high status kinship and low status kinship respectively. Now, these styles of behaviour may be called roles because they are not fixed and immutable aspects of the life of a given individual. On the disappearance of a control animal another male typically adopts the same role. A male rising or falling in a dominance hierarchy changes his behaviour. Fallen males may, however, march out of the group with affiliated females and establish a new group in which they may adopt the role of a central animal. Isolate males may enter a group and form a central sub-group, one of them perhaps being the control.

While not every conceivable role is necessarily present in any given group there is an overall consistency that makes this approach meaningful. Certain roles are usually characteristic of some kinds of groups and not of others — the harem "overlord" occurs in Hamadryas, Patas and Gelada groups but not in other baboons or among macaques.

The behaviour characteristic of a particular role is not the "property" of the individual playing it. Such behaviour is not fixed by conditioning so that the individual remains forever the same. Physiological and social changes impel behavioural shifts so that in a lifetime individuals may play many roles in their social structures. Social mobility in the sense of role changing is an important attribute of primate groups. It seems more characteristic of males than of females. Young and subadult males go solitarily, live peripherally or in all-male groups or shift from one reproductive unit to another. Females are more loyal to their natal group providing the more fixed social positions around which males revolve. This mobility is an expression of a set of tensions or forces characteristic of group life. At least three sets of factors interact in this social sorting process (Crook 1970c). These are: (i) the maturing and aging of individual group members, (ii) the growth and splitting of groups, and (iii) intra-group competition for environmental and social commodities together with the affiliative cooperation and subterfuge that this entails.

The third set is the most critical at any one time, the other factors acting over longer periodicities. There appear then at any one time to be two opposed social processes operating within a baboon or macaque troop. One consists of the assertive freely expressed utilization of available physical and social commodities by high status animals which has the effect of constraining the behaviour of juveniles and subordinates. The other consists of the adoption of behavioural subterfuge by certain subordinates whereby such behavioural constraint can be avoided (e.g., by temporary solitary living or by splitting up the group into branch groups permitting a greater freedom of expression to the new leaders).

This "subterfuge" is presumably not conscious or deliberate. It appears to arise from the need of certain subordinates to free themselves from the behavioural constraints to which they have been exposed. There are in fact relatively few known "routes" along which such an aspirant may move. These are: (a) temporary solitarization, followed by take-over of a small branch group; (b) the use of an affiliated companion to ensure the rise of a dominant animal into a higher or even a centrally located position; and (c) the care given by "aunts" and "uncles" to infants and babies as a means of entering high-status sub-groups and so to affiliate with them.

Examples of the last case are particularly interesting. Itani (1959) showed that adult male macaques become interested in juvenile animals when the latter are 1 year old. At this time females are in the birth season and cease protecting the young of a previous year. The males appear to show interest in young animals only in this period. Their behaviour consists of hugging and sitting with the infant, accompanying it on the move, protecting it from other monkeys and dangerous situations and grooming it. Although Itani found no relationship between dominance as such and the frequency of child care he did find the behaviour to be especially developed in middle-classed animals of the caste structured troops. The behaviour indeed seemed pronounced in animals that were sociable, not aggressive and oriented towards high caste animals. Itani suggests that certain monkeys by showing care to high caste children succeed in being tolerated by leaders and their affiliated females.

Such animals may therefore rise in rank. Protectors usually behave in a mild manner which may facilitate both their association with infants and their rise in rank. It is not clear, however, whether the effect on rank lasts for more than the birth season. Itani gives no details of permanent changes in social structure so produced. The adult males appear to protect 1-year-old males and females equally but more second-year females than males are protected owing to the social peripheralization of young males. Protected 2-year-olds are moreover often poorly grown individuals that were protected by the same male in the previous year. Protection has a further interesting effect in that adults more readily learn to investigate the new foods a progressive infant has discovered than would otherwise be the case.

Itani found cases of male care common only in three of eighteen troops investigated. It occurred very rarely in seven others and was not observed in the remaining eight. The behaviour thus appears to be a local cultural phenomenon. It undoubtedly provides additional chances of survival for young animals likely to be relatives, increases the rate of spread of new patterns of behaviour in the group and, since its frequency differs between troops, may increase the reproductive success of some monkey groups over others.

Recent observations on the Barbary Macaque (*Macaca sylvana*) in the Moyen Atlas of Morocco show that adult males in wild groups show an extraordinary amount of interest in young babies of the year. An interest in babies so young was relatively uncommon in Itani's study. Furthermore, the male Barbary Macaque does not seem, on present evidence, to limit his interest to a particular infant. He appears to appropriate babies from females in the group and to groom and care for them for short periods usually under 15 min. duration. Babies may, moreover, move away from their mothers to accompany males, commonly riding off on their backs. Babies also move from one male to another. Males carrying babies frequently approach males without them in such a way as to encourage the approached animal to engage in a mutual grooming session with the baby as target. In some of these sequences babies are presented while on the back of the approacher, the animal turning its rear towards the other male (as it would do in sexual presenting) as it does so. In one case the approached male was seen to mount and thrust an animal that had done this. It appears too, that most males approaching with babies are relatively juvenile animals. It looks as if relatively subordinate animals are using the babies in some way to improve their relations with higher ranking males (Deag and Crook, 1970).

So far too little is known about the social structure of the Barbary Macaque to show whether there is a caste system comparable to that described from Japan. Whether males are using the babies as a means to increase their social standing and, hence, their freedom to behave without the constraints imposed by low rank, remains unknown. It does seem, however, that in both species not only does the male's behaviour increase the chances of survival for young animals but it appears to be closely related to the structure of the monkey group and the patterns of constraints regulating social mobility of individuals within it. We do not know yet whether the behaviour of the Barbary Macaque male is restricted, as it is with the Japanese Macaques, to certain troops and localities or whether it is a general phenomenon found throughout the species.

In all these cases individuals closely affiliated to others can combine cooperatively to bring about the liberating effect, an effect which, furthermore, in times of shortage (food, females) would ensure an increased probability of survival and/or reproductive success in addition to the psychological freedom from the effects of "stress" that undoubtedly accompany in some degree a continuing social restraint on individual behavioural expression. Similarly, by enlisting the cooperation of affiliated animals, often relatives, in their "control" behaviour, animals that are already established in social positions of high rank will be able to maintain such positions, and the access to commodities it provides.

It seems, then, that cooperative behaviour of the high order found in macaque and baboon groups has arisen within the context of competition for access to both physical and social commodities (Crook 1970c discusses). Both direct affiliative behaviour and indirect affiliation, for example through a common interest in child care, provide the basis for common action. As we have seen such cooperation provides numerous advantages for the participants, and not only for the most dominant animals among them. It also provides the behavioural basis for the complex class-structured society of these animals with its tolerance of individual mobility between roles.

Finally, the stable social structure maintained by a powerful clique around a control animal seems to provide the optimum circumstances for maternal security and child rearing. Females form the more cohesive elements of primate groups and, as a consequence of their affiliative relations and kinships links, may play a much greater role in determining who emerges as "control" than is at present known. Males by contrast, subject to the full force of social competition, are the more mobile animals transferring themselves, as recent research shows, quite frequently from one group to another.

CONCLUSIONS: SIGNIFICANCE FOR HUMAN ETHOLOGY

In this paper I have tried to present a conspectus of the current state of social studies in ethology. Much more could of course have been added and not all would adopt the particular orientation I have used. It is my belief that an effective ethology of man must be based primarily on social considerations. Such a basis may possibly be supplied by a viewpoint such as I have adopted here. Often inherent in this approach is the use of social psychological ideas as a useful adjunct to ethological analysis. This mutuality between two disciplines is likely to develop quickly probably to the benefit of both.

There seems to be a problem in differentiating "human ethology" from the already existing behavioural sciences of man although Waxweiler had in 1906 already made clear that ethology as a broad biopsychological discipline included sociology and its derivatives as a special branch subject focussing upon the problems of man. Today it seems that by human ethology we designate not so much a subject as a way of approaching human behaviour viewing the species as a member of the animal kingdom rather than as a separate and peculiar phenomenon. This I believe to be most healthy so long as one remembers that man is indeed a peculiar phenomenon and that mere biological reductionism, a "nothing butism" as Julian Huxley used to say, will get us nowhere.

An attempt to apply classical ethology to man would suggest that an ethological version of kinesics and proxemics was in the offing together with an elaboration of evolutionary theories regarding the adaptive significance of breasts and beards. Although interesting this would seem too curtailed an approach.

My feeling is that human ethology should comprise the whole behavioural biology of man. Such a science, it seems to me, must focus its attention on at least the following problems:

(1) The evolution and cultural history of the basic human grouping structures, families, all-male congregations, etc., and their origins in non-human primate grouping patterns (Reynolds 1966, 1968).

(2) The relations between human individual behaviour, group composition and population density. How far do the problems of rodent stress-physiology apply also to man?

(3) Non-verbal communication through faceal and postural expression. The evolution of emotional expression in man.

(4) The functioning of non-verbal communication in the control of affect in small groups, the rules of interpersonal ethology in companionship and courtship, the uses of ethology in sensitivity training, control of interpersonal aggression, etc. (Goffman 1963; Argyle 1967).

A programme as broad as this needs a basis within the framework of a biological social ethology such as I have tried to outline in this paper and which must include at least the three perspectives discussed. The scope is clearly very wide with links to anthropology, demography, stress physiology, ecology and particularly social psychology and psychiatry. To obtain workers equipped for research in this area requires an emphasis on training and a focus on such problems within the university syllabus. The day of routine study patterned by the ethological traditions of the past is over. We need fresh orientations and to find them social ethologists must go looking in new places along unexplored paths.

Summary

1. A major development within the ethology of the last decade focusses attention upon the relations between social behaviour, ecology and population dynamics. This field may be termed "social ethology" following an early but neglected usage of the term by Waxweiler at the start of the century.

2. Contemporary social ethology comprises three interdependent perspec

tives, socio-ecology, socio-demography and the study of social processes within natural and experimental groups.

3. In socio-ecology recent studies reveal that close correlations exist between the forms of avian and mammalian social organizations and their respective ecological niches. In particular the adaptive significance of certain mammalian societies comprising, on the one hand, multi-male reproductive units and, on the other, those made up of one-male and all-male units is discussed and explanatory hypotheses derived from primate and ungulate data briefly considered.

4. In socio-demography research suggests that socially mediated mortality is of greater significance in the density-dependent control of bird and mammal numbers than had formerly been thought. The relations between ecological factors extrinsic to and social factors intrinsic to a social organization may be modelled in the form of open adaptive cybernetic systems rather than expressed in terms of analogies to closed mechanical or physiological systems.

5. Studies of social processes in non-human primate groups suggest that some form of role analysis may prove heuristic. The relations between dominance status, affiliative and kinship relations, social subterfuge and competition-contingent cooperation are discussed in an attempt to outline the dynamics of social change within relatively stable group structures.

6. "Human ethology" seems at present to lack adequate definition as an academic discipline. Social ethology may provide the essential biological basis for future research in this area.

Acknowledgments

Revised text of an address to the plenary session on Social Organization, XIth International Ethology Conference, Rennes, France, 1969. Text prepared during the tenure of a Fellowship at the Center for Advanced Study in the Behavioral Sciences, Stanford, California, 1968 to 1969.

I am most grateful for advice and comment on the text of this paper, whether in whole or part, kindly given me by Ray Birdwhistell, L. Poser, Kai Erikson, John Goss-Custard, Robert Hinde and Henri Tajfel.

References

Aldrich-Blake, F.P.G. 1970. Problems of social structure in forest monkeys. In Social Behaviour in Birds and Mammals, ed. J.H. Crook. London: Academic Press.
Allee, W.C. 1938. Cooperation Among Animals. New York: H. Schuman.
Argyle, M. 1967. The Psychology of Interpersonal Behaviour. London: Pelican.
Ashmole, N.P. 1961. The biology of certain terns. D. Phil. Thesis, Oxford.
_____. 1963. The regulation of numbers of tropical oceanic birds. Ibis, 103b:458-73.
Bernstein, I.S. 1966. Analysis of a key role in a capuchin (Cebus albifrons) group. Tulane Stud. Zool. 13:49-54.
Bernstein, I.S., and Sharpe, L.G. 1966. Social roles in a rhesus monkey group. Behaviour 26:91-104.

Buckley, W. 1967. Sociology and Modern Systems Theory. New York: Prentice-Hall.

Chitty, D. 1967. What regulates bird populations? Ecology 48:698-701.

Coulson, J.C. 1968. Differences in the quality of birds nesting in centre and on the edges of a colony. Nature, Lond. 217:478-79.

Crook, J.H. 1964. The evolution of social organisation and visual communication in the weaver bird (Ploceinae). Behaviour, Suppl. 10.

———. 1965. The adaptive significance of avian social organisations. Symp. zool. Soc. Lond. 14:181-218.

———. 1966. Gelada baboon herd structure and movement, a comparative report. Symp. zool. Soc. Lond. 18:237-58.

———. 1970a. Social behaviour and ethology. In Social Behaviour in Birds and Mammals, ed. J.H. Crook. London: Academic Press.

———. 1970b. The socio-ecology of primates. In Social Behaviour in Birds and Mammals, ed. J.H. Crook. London: Academic Press.

———. 1970c. Sources of cooperation in animals and man. In Man and Beast, Comparative Social Behaviour. IIIrd International Symposium, Smithsonian Institution, 1969.

Crook, J.H., and Aldrich-Blake, P. 1968. Ecological and behavioural contrasts between sympatric ground dwelling primates in Ethiopia. Folia primat. 8:192-227.

Crook, J.H., and Butterfield, P.A. 1970. Gender role in the social system of Quelea. In Social Behaviour in Birds and Mammals, ed. J.H. Crook. London: Academic Press.

Cullen, E. 1957. Adaptations in the Kittiwake to cliff-nesting. Ibis 99:275-302.

Cullen, J.M. 1960. Some adaptations in the nesting behaviour of terns. Proc. XII Int. Orn. Congr., Helsinki, 1958, pp. 153-57.

Deag, J., and Crook, J.H. 1970. Social behaviour and ecology of the wild Barbary Macaque (Macaca sylvana [L]). In preparation.

Delius, J.D. 1965. A population study of skylarks, Alauda arvensis. Ibis 107:465-92.

Espinas, A. 1878. Des Societes Animales. Paris: Baillière.

Estes, R.D. 1966. Behaviour and life history of the Wildebeest (Connochaetes taurinus Burchell). Nature, Lond. 212:999-1000.

Gartlan, J.S. 1968. Structure and function in primate society. Folio primat. 8:89-120.

Goffman, E. 1963. Behavior in Public Places. New York: Free Press.

Hall, K.R.L., and DeVore, I. 1965. Baboon social behavior. In Primate Behavior, ed. I. DeVore, pp. 53-100. New York: Holt, Rinehart & Winston.

Hinde, R.A. 1966. Animal behaviour. In A Synthesis of Ethology and Comparative Psychology. New York: McGraw-Hill.

Huxley, J.S. 1959. Clades and grades. Systematics Ass. Publ. No. 3. Function and Taxonomic Importance, pp. 21-22.

Immelmann, K. 1962. Beiträge zur Biologie und Ethologie australischer Prachtfinken (Spermestidae). Zool. Jb. (Syst.) 90:1-196.

———. 1967. Verhaltensökalogische Studien an afrikanischen und australischen Estildidae. Zool. Jb. (Syst.) 94:1-67.

Itani, J. 1959. Paternal care in the Wild Japanese Monkey. Macaca fuscata. J. Primat. 2:61-93.

Jarman, P. 1968. The effect of the creation of Lake Kariba upon the terrestrial ecology of the Middle Zambezi valley, with particular reference to the large mammals. Ph.D. Thesis, Manchester University.

Jenkins, D.; Watson, A.; and Miller, G.R. 1963. Population studies on Red Grouse, Lagopus I. scoticus. J. anim. Ecol. 36:97-122.

Klopfer, P.H. 1962. Behavioral Aspects of Ecology. Englewood Cliffs, N.J.: Prentice-Hall.

Kluyver, H.N., and Tinbergen, L. 1953. Territory and the regulation of density in titmice. Arch. neerl. Zool. 10:266-87.

Kropotkin, P. 1914. Mutual Aid, a Factor in Evolution. New York: A.A. Knopf.

Kummer, H. 1968. Social organisation of Hamadryas baboons. Bibliotheca Primatologica, No. 6.

Lack, D. 1954. The Natural Regulation of Animal Numbers. Oxford: Clarendon Press.
_____. 1966. Population Studies of Birds. Oxford: Clarendon Press.
_____. 1968. Ecological Adaptations for Breeding in Birds. London: Methuen & Co.
Lockie, J. 1956. Winter fighting in feeding flocks of Rooks, Jackdaws and Carrion Crows. Bird Study 3:180-90.
Murton, R.K.; Isaacson, A.T.; and Westwood, K.T. 1964. The relationships between woodpigeons and their clover food supply and the mechanism of population control. J. appl. Ecol. 3:55-96.
Nadel, S.F. 1957. The Theory of Social Structure. Glencoe, Ill.: Free Press.
Nelson, J.B. 1967. Etho-ecological adaptations in the Great Frigate-bird. Nature, Lond. 214:318.
Orians, G.H. 1961. The ecology of blackbird (*Agelaius*) social systems. Ecol. Monogr. 31:285-312.
Patterson, I.J. 1965. Timing and spacing of broods in the Black Headed Gull (*Larus ridibundus*). Ibis 107:433-59.
Petrucci, R. 1906. Origine polyphylétique, homotypie et non-comparabilité direct des sociétés animales. Travaux de l'Institut de Sociologie. Notes et Memoires, 7. Bruxelles: Instituts Solvay.
Pitelka, T.A. 1942. Territoriality and related problems in North American humming-birds. Condor 44:189-204.
Reynolds, V. 1966. Open groups in hominid evolution. Man 1:441-52.
_____. 1968. Kinship and the family in monkeys, apes and man. Man 2:209-23.
Rowell, T.E. 1966. Hierarchy in the organization of a captive baboon group. Anim. Behav. 14:420-43.
Sarbin, T.R. 1959. Role theory. In Handbook of Social Psychology, ed. G. Lindzey. Cambridge, Mass.: Addison-Wesley.
Solomon, M.E. 1964. Analysis of procedures involved in the natural control of insects. In Advances in Ecological Research, ed. J.B. Crag, Vol. 2. London and New York: Academic Press.
Stonehouse, B. 1960. The King Penguin *Aptenodytes pategonica* of South Georgia. I. Breeding behaviour and development. Sci. Rep. Falkland Is. Depend. Surv. 23:1-181.
Tinbergen, N. 1964. On adaptive radiation in Gulls (*Tribe Larini*). Zool. Meded. 39: 209-23.
_____. 1967. Adaptive features of the Black Headed Gull (*Larus ridibundus* L.). Proc. XIV Int. Orn. Congr., Oxford, 1966, pp. 43-59. Oxford: Blackwell.
Tompa, F.S. 1964. Factors determining the numbers of Song Sparrows *Melospiza melodia* (Wilson) on Mandarte Island, B.C. Canada. Acta Zool. fenn. 109:1-68.
Ward, P. 1965. Seasonal changes in the sex ratio of *Quelea quelea* (Ploceinae). Ibis 107:397-99.
Waxweiler, E. 1906. Esquisse d'une Sociologie. Travaux de l'Institut de Sociologie. Notes et Memoires, 3. Bruxelles: Instituts Solvay.
Wilson, A.P. 1968. Social behaviour of free-ranging rhesus monkeys with an emphasis on aggression. Ph.D. Thesis, University of California, Berkeley.
Wynne-Edwards, V.C. 1962. Animal Dispersion in Relation to Social Behaviour. Edinburgh: Oliver & Boyd.

Seven

Stephen Gartlan's review of the concept of social dominance in primate society falls in the purview of the third of the three perspectives which Crook sees as making up social ethology. The concept of social dominance as a major structuring mechanism in nonhuman primate groups (the *main axis* of social organization, as it has been frequently and summarily described) has had a profound influence on the goals and methods of primate field studies, not to mention the interpretations drawn from them. Gartlan points out that unitary theories of social structure and function which posit the sexual bond as the "basic cohesive element of primate congregations" and social dominance as the main determiner of "the structure of the congregations" are derived mainly from observations of captive animals in laboratory and zoo settings. Such theories have proved inadequate in analyses of the social organization of natural, free-ranging primate groups, observed differences in which appear to be as much a product of habitat as of genetic inheritance and specific variation. Gartlan examines in detail "the shortcomings of social dominance theory as ... a descriptive and an analytical concept" and suggests that differences in primate social patterns might be better understood and a more adequate ground laid for theories of primate behavioral evolution if social structure were to be described in terms of role differentiation.

Structure and Function in Primate Society

BY J. STEPHEN GARTLAN

INTRODUCTION

The number of field studies of free-living primates has increased greatly over the last decade and there is now a considerable body of knowledge concerning the distribution and social structure of many different species. The origin of interest in this area of research lies to a large extent in pioneer investigations of captive animals carried out mainly by comparative psychologists such as Köhler (1925) and Yerkes and Yerkes (1929). The orientation of these early studies, however, was chiefly towards the comparison of mental processes and the phenomena of learning and not towards problems of social structure and behaviour. Eventually, therefore, when attempts were made to construct theories explaining primate congregations and their structure, it was perhaps inevitable that they were based in mentalistic theories of proximate causation and only loosely related to selection theory and evolution. Theories important in influencing later field studies were derived mainly from studies of caged and captive individuals, often living as individuals or as pairs. Under these conditions the problem of the adaptive significance of group structure did not arise, except that captivity, which is essentially an impoverished habitat, often resulted in the breakdown of normal social patterns and in a high incidence of disease and death.

The resulting two unitary motivational theories which have influenced the majority of all subsequent field and laboratory studies of primates are, (1) that the "sexual bond" is the basic cohesive element of primate congregations, and (2) that "social dominance" determines the structure of the congregations. Perhaps the most explicit and succinct statement of the supposed function of these two mechanisms was that of Zuckerman (1932): "The main factor that determines social groupings in sub-human primates is sexual attraction. Females attract males and males attract females. The limit to the number of females held by any single male is determined by his degree of dominance." The first of these theories has been the subject of much

Reprinted with permission from *Folia Primatologica*, Vol. 8, no. 2, 1968, pp. 89-120.

recent discussion (c.f. Lancaster and Lee [1965]) and will not be considered in detail in this analysis which is concerned mainly with the theory of social dominance.

In a review of avian social organisation, Crook (1965) noted that changes in social structure are best considered in terms of shifts within a balanced drive relationship. Selection pressures do not act on social organisation as such, but rather on particular tendencies — such as nest site attachement, escape or sex — a shift in any one of which may result in changes in the social system as a whole. Whilst it is theoretically possible, therefore, for primate social groups to be determined and structured by variations in the strength of two unitary parameters, multifactorial determination seems phylogenetically more probable. Clearly, however, this approach demands analysis of the behaviour patterns and ecological factors important in the structuring of the society. These factors are many and diverse, only one of which may be, from time to time, the necessity to limit or reduce population size relative to available food resources. Yet the social dominance hierarchy, originating from captive studies, and considered by some population ecologists to have the primary function of relating population size to available resources, has often been considered as the sole primate social structuring mechanism.

Whilst the results of hierarchies in conditions of food privation are not generally disputed, their evolutionary significance is the subject of considerable debate. Lack (1966) has argued that hierarchies may be explained in terms of natural selection, so that a hierarchical social structure would result from dyadic interactions in the privation situation, arising therefore as a response to scarce food resources and not as an anticipation of them. The contrary view is taken by Wynne-Edwards (1962) who considers the hierarchy to be preadaptive for conditions of privation and to arise through group selection. According to this theory (Wynne-Edwards, p. 141), ". . . what is passed from parents to offspring is the mechanism, in each individual, to respond correctly in the interests of the community — not in their own individual interests — in every one of a wide range of social situations. " At this point it seems necessary to distinguish between mechanisms of population dispersion and intra-group social structure. Wynne-Edwards' view that hierarchies distribute the population through the environment leads to the construction of social dominance hierarchies which are unrelated to intra-group social structure and which are in fact a direct function of the distance of a breeding territory from an artificially introduced, highly concentrated food source. Thus Brian (1949) in a study of the great tit, *Parus major*, noted that a straight line dominance order existed among seven males, although this was broken at one point by a triangular relationship. This hierarchical structure was however determined at a single feeding station. Each bird when not feeding at the station tended to return to its breeding territory. Brian noted that the position of the bird in the dominance order corresponded with the distance of its territory from the feeding place; the further away a bird's territory, the lower its status.

A similar confusion arises in some studies of mammalian populations. In an attempt to describe the woodland social structure of grey squirrels, *Sciurus carolinensis*, Taylor (1966) although noting that each animal had a particular territory, nevertheless established only a single feeding station for the determination of social rank. Consequently it was not surprising that the dominant animals were those with territories nearest the feeding station, while the subordinates were those having to cross most foreign territories on the journey to the food hopper. Such experimental techniques fail because they impose social models on what are not social units.

Primate social units are usually relatively permanent, with feeding occurring only within the home range or territory. There is therefore less danger of confusing territorial behaviour with intra-group social interactions, although one possible source of error is the exchange of adult males between social groups. This has been noted in several species, notably *Cercopithecus aethiops* (Gartlan [1966]), *Papio anubis* (Rowell, pers. comm.), *Gorilla gorilla beringei* (Schaller [1963]) and probably also in *Macaca fuscata* (Hazama [1964]). Aggression against any given adult male could possibly result from immigration in the wild, but unless the history of the group in question is known, it is impossible to control for this phenomenon.

Appropriately applied, social dominance theory attempts to describe intra-group social structure. However, the ordering of individuals into a linear hierarchy depends, both logically and in terms of selection theory, on an underlying continuum of rank-order criteria. Kummer (1957) in a study of *Papio hamadryas* noted that aggressive patterns, although good indicators of rank in adult females, failed to differentiate rank in two-to-three-year-olds and were inappropriate in one-year-olds, which show play rather than aggression. This illustrates a methodological problem; differentiation of function within primate groups is generally such as to be outside the scope of analysis in unitary terms.

A further source of confusion arises from those behaviour patterns which are traditionally associated with dominance — aggression, copulation and access to food — and which are indeed often used as indices of social dominance. These behaviour patterns often show no correlation with one another. Kummer (1957) for example noted that the normal rank of adult males based on aggressive interactions was subject to inversion in the field of sexual behaviour. Hall and DeVore (1965) record that in a baboon group in Nairobi Park, Kenya, an old male with teeth worn down to the gums and individually the least dominant animal was nevertheless second only to the most dominant male in number of copulations completed at the time of maximum sexual swelling of the female. Southwick, Beg and Siddiqi (1965) record consort relationships of subordinate males with oestrus females in wild populations of *Macaca mulatta*. Jay (1965) noted that the number of completed copulations in langur, *Presbytis entellus* groups did not always correspond with the dominance rank of the male. Jolly (1967) records a linear dominance hierarchy amongst males of *Lemur catta* in the wild. This hierarchy, based on aggressive encounters, was not correlated with mating success. During the extremely short mating period the aggression hierarchy broke down and fights over sexually receptive females were common, yet the fourth-ranking out of five adult males in one group won access to females and copulated most. Nevertheless, after the end of the mating season this animal was observed still to occupy his previous low-raking position in the group.

The frequent use of the number of copulations as an index of social dominance has characterized many primate field studies, even though conflicting results are reported in early avian studies. Although Murchison (1935) and Skard (1937) both noted a positive correlation between social rank and frequency of mating in flocks of hens, *Gallus domesticus,* completely opposite results were obtained by Allee (1952). This technique of frequency of mating as an index of social dominance carries evolutionary implications which have rarely been questioned. If social dominance were a genetic character or complex of characters, heritable in the same manner as coat-colour or other physical characteristics, it would follow that the exclusion of non-dominant males from an effective breeding role would constitute intense selection for social dominance. Long-term studies in Japan and on Cayo Santiago do not support this contention. On the contrary, Kawai (1958) and Kawamura (1958) on *Macaca fuscata* and Sade (1967) on *M. mulatta* have indicated that the social rank of the infant is directly dependent on that of the mother and subject to cultural influences and learning. The most dominant male, in other words, if it succeeds in fertilising all the adult females in a group, will father infants of which the social ranks will be directly correlated with the ranks of the mothers and not related to patterns of Mendelian segregation. These findings indicate strongly that within the free-ranging group social dominance is not a unitary character or a "property" of individuals, but that it is a role or a series of roles, and that the relationship with sexual behaviour is incidental rather than causal.

The comparative rarity of linear hierarchies in wild populations and the more common finding of triangular relationships, reversals and "central hierarchies" is difficult to explain in terms of inherited tendencies or a unitary structuring mechanism, but more easily explicable as the result of learning in the dyadic situation. This is confirmed by experimental studies which have often failed to establish a strong correlation between position in the social hierarchy and such characteristics as age, sex, size, physical strength and the ability to learn mazes. Field studies have often reported phenomena which can only be meaningfully explained in terms of learning. Thus Simonds (1965) in a study of *Macaca radiata* recorded that a very old, canine-less adult male, who was actually subordinate to the central males and who was not very active in threat sequences or other agonistic behaviour, nevertheless had the "effect" of a dominant male inhibiting or terminating agonistic encounters between other group members.

The paradoxical and confused results of attempts to apply unitary structuring concepts to such complex social groups as those of the primates necessitates a critical reappraisal of the usefulness of the concept as an analytical tool in field studies. The wider question of the adaptive significance and evolution of social structure has also received new importance from the results of several recent field studies recording variability in intra-specific group structure frequently exceeding that between species. The most significant studies in this respect have been those by Jay (in press) on different group sizes, social structure and behaviour of *Macaca mulatta* in the different habitats of forest, roadside and city; the different social structures of *Presbytis entellus* in the evergreen forests of Orcha in Madhya Pradesh (Jay [1965]) compared with that of the same species in the seasonally arid areas of Dharwar in Mysore (Sugiyama [1965;1967]), and differences in social structure and behaviour of *Cercopithecus aethiops* described by Gartlan and Brain (in press) and Struhsaker (pers. comm.) which were correlated with different ecological conditions.

Other influences requiring critical reappraisal of the concept have been attempts to generalise from studies of primates to man. Thus Maslow, Rand and Newman (1961) stated that " . . . the dominance-sexual fusion is a lower evolutionary development than the differentiation of sex from dominance and parallels our suspicion that such a differentiation in the human being may be a correlate or epiphenomenon of greater psychological maturity or development." Such theories rely heavily on social dominance as a genetically heritable, ubiquitous structuring mechanism in non-human primates. The important implications for human behaviour require that the biological assumptions of the theory be closely examined.

RESULTS

UNITARY THEORIES OF COHESION AND STRUCTURE

1. *Sexual bonding.* The theory of the sexual bond as the cohesive factor in primate congregations was first advanced by Zuckerman (1932). This hypothesis was stated explicitly and most subsequent field studies have attempted to obtain data on the periodicity of mating and the presence or absence of birth seasons as a critical test of the theory. Data available up to 1965 were summarised by Lancaster and Lee (1965) who concluded that constant sexual attraction could not be the sole basis for the persistent groupings of primates. There can be no doubt, however, that sexual motivation is capable of providing integrative cohesion in a multifactorially determined social structure. Primate social units seem in the majority of cases to be stable over time in spite of the fact that adults, especially females, do not spend a great deal of time in sexual activity. Rowell (1967) noted that the adult female baboon is usually either pregnant or lactating and not sexually receptive. Sade (1964) showed that the adult male rhesus macaque, *Macaca mulatta* undergoes a cyclic change in testi size and spermatogenesis, indicating seasonal sexual receptivity; a finding which also applies apparently to an exotic population of *Cercopithecus aethiops sabaeus* on St. Kitts Island, in the West Indies (Sade and Hildreth [1965]).

The persistence of the social group in primates contrasts with the position in many avian and mammalian species in which the social group is an exclusively reproductive unit, set up at the beginning of the breeding season and dispersing after the rearing of the young. There is no evidence of a breakdown in the group following the rearing of infants in primate species having a restricted birth season. Jolly (1967) records a reduction in total social inter-actions in *Lemur catta* groups following the extremely short breeding season, but this was not associated with breakdown of the groups. The transfer of adult males between groups of *Cercopithecus aethiops* was not a symptom of mass social breakdown, but was a rare and individual occurrence. Adult males tended to occur away from other group members and were the most frequent victims of aggressive chases. In general it was concluded (Gartlan [1966]) that the enduring social unit in this case was the adult female — juvenile — infant complex. It should be noted also that Washburn and DeVore (1961) record sexual behaviour as a disruptive factor on baboon social organisation. Such findings weaken further the unitary sexual bond theory.

Many field studies have indicated that although sexual behaviour does have cohesive results, other behaviour patterns may play equal or even more pro-nounced roles in group cohesion. Other cohesive factors include the birth of an infant, the enduring mother-infant relationship and the intrusion into the group's territory by an alien group. Furthermore, emotional responses to acute stress situations in young rhesus monkeys are less likely to occur and less severe when they do occur if the experimental subject is accompanied by a familiar partner or even an unfamilar age-peer (Mason [1960]). This indi-cates a further possible cohesive factor and there are probably many others awaiting empirical determination.

2. *Social dominance.* The theory of social dominance, unlike that of the sexual bond, cannot be attributed to any single worker. However, several of the most important influences on the development of the theory have been the work of William McDougall (1908), T. Schjelderup-Ebbe (1931) and A. H. Maslow (1936).

McDougall's social psychology postulated a multiplicity of "social instincts." There was a tendency in this scheme for every distinguishable social behaviour pattern to be ascribed to the operation of a unitary "social instinct." Examples include the instincts of "self-assertion," "self-abasement" etc., many of which had a significant influence on the development of social dominance theory, and particularly in the postulation of a distinct dominance drive. This theoretical background was reinforced by the experimental studies and derived theories of later workers. Schjelderup-Ebbe (1931) for example, working on the peck-order in bird flocks, interpreted his results within a metaphysical theory of "despotism" which was the "basic idea of the world."

Social dominance has been used as both a descriptive and an analytical concept in primate biology, with many different meanings and shades of meaning (Gartlan [1964]). The result of this has been that there is often ambiguity as to which precise behaviour pattern the term refers. It has been used as a synonym for aggression, for sexual, territorial or other behaviour patterns, and as a proposed mechanism of population dispersion, often with-out operational definition. Occasionally even the simplest operational definition has been found inadequate and incompatible with hierarchy theory, but even this has led only infrequently to redefinition of the concept. For example, Chance (1956) stated that "Dominance is usually defined as priority of access to a need-satisfying object . . . but, since in this instance no priority of feeding was shown, it was in the relations of the animals to each other that the order of rank was made manifest." The development of the theory of social dominance as the normal structuring mechanism of primate societies was permitted in part by the observation of similar rigidities of social behaviour in different species of primate under captive conditions, and the postulation that dominance influenced most, if not all, aspects of social behaviour. For example, Chance (1956) concluded that "The dominance relations in Macaques as well as Baboons . . . are, therefore, conspicuous by being constantly present and influencing every aspect of behaviour. Dominance gives priority of access. The probability, therefore, of access to an incentive appropriate to a monkey's activated drive may be expected to depend on the degree of dominance exerted by that monkey." This quotation is remarkable in two main respects. Firstly it contradicts the first quotation from Chance (1956) where dominance was observed not to influence every aspect of behaviour. Secondly it demonstrates the circular argument common in this field: priority of access defines dominance, dominance gives priority of access. This results in a confusion between force and criteria; the fallacy that one is making a statement of a different order of complexity and identifying causal factors in saying that dominance gives priority of access. All that may be properly inferred is that dominance is priority of access.

There is evidence from many of the early studies that these populations were under severe social stress; a phenomenon which will be described in the appropriate section, but which has similar and well-defined results in many mammalian species. The behaviour exhibited in captivity and which permitted the construction of the social dominance hypothesis as a descriptive and explanatory concept may therefore be only indirectly related to the structure and behaviour of groups living under adequate environmental conditions in a habitat to which they are adapted.

Conclusions about the ubiquity of social dominance as a social structuring mechanism were also permitted by many experiments which were inadequate in design and experimental criteria. In Maslow's investigation (1936) for example, groups in which the comparative levels of dominance were assessed comprised different genera and species of primate as well as different age-sex classes, and included in one instance a Coati, *Nasua* sp. Similar limitations applied to the investigation by Maslow and Flanzbaum (1936) in which animals were housed individually and tested in pairs which were "equated for weight, age, sex and *usually species.*"

Pioneer workers going out into the field for the first time were inevitably influenced by the findings and theories of the comparative psychologists, and even if not expecting social uniformity, were hardly prepared conceptually for the variability which was discovered. As it became clearer that a simple, linear hierarchy was inadequate to deal with the observed social complexity, a tendency appeared for the concept to be altered, not by critical reappraisal, but by ad hoc modification. Thus DeVore (1965) postulates a central hierarchy, "If one of the males in the central hierarchy is threatened, the other males tend to back him up. Thus this central group of males outranks individuals, even though a young male outside the central hierarchy might be able to defeat any member of it separately." Statements of this kind, it is important to note, result from the clear failure of the traditional social dominance theory to fit field data. And although this represents a fundamental revision of the concept in that individual, egocentric motivation is subordinated to, and to some extent replaced by, social co-operation, yet the tendency to adhere to a hierarchical conceptual framework remains, although, as in this case, it seems hardly relevant.

SOCIAL CHANGE IN CAPTIVITY AS REVEALED BY FIELD STUDIES

The analysis of variability in social structure elucidates those factors important in determining social characteristics; hence the particular value of field studies (1) of species that have previously only been studied in captivity, and (2) of wild populations of single species in different habitats. The first point is well illustrated by the pioneering studies of K. R. L. Hall on *Erythrocebus patas.* Hall (1965) failed to see any actual fighting in the field, and aggression was very rare. He observed a total of only 49 encounters during a total of 627 observation hours. Of these 49 incidents only 10 resulted in actual physical contact and of these only 2 led to brief biting, with no visible damage inflicted in either case. Of the 49 cases, adult females were involved as aggressors in 30, and Hall notes that this was almost certainly an underestimate as the sexually indeterminate adults were almost certainly females. Of these 30 cases, 11 were directed at the adult male of the group. In captivity, on the other hand, Hall, Boelkins and Goswell (1965) record aggressive threat in *E. patas* as being fairly common, indicating a marked increase under these conditions.

The comparative rarity of aggression in wild populations of *Papio ursinus* was also noted by Hall (1962). Of 33 recorded instances of chasing and chastising by adult males, only one involved another adult male. No serious injury resulted from these incidents. Male 1 of S group was the aggressor against females on 21 occasions and juveniles on 8 occasions. These 29 incidents were spread over 190 observation hours, about one incident for each 6½ h of daylight.

Gartlan (1966) noted that in two captive groups of *Cercopithecus aethiops* the incidence of aggression was considerably higher than in a wild population. The mean rate in a captive group at Bristol was almost 2 incidents per hour, compared with 0.28 incidents an hour on Lolui Island, Uganda. The figures for the captive group are the mean of 45 h of observation, whilst those for the wild population represent over 1200 observation hours. Correlated with this was an increase in "social monitoring" in the captive situation. This behaviour pattern, which was mainly performed by juvenile animals, was measured by the number of times an individual in the group clearly took note of the position of another individual in the group by turning the head or body, or by pronounced eye movements. Precise measures were not obtained in the field, but in the captive situation a juvenile male was observed to make 420 such movements during 10 observation hours, certainly many more than were usual in the wild, by about 10 times.

In a precise and detailed study of *Papio anubis* in captivity, conceived as a control to studies of the species in the wild, Rowell (1966) concluded that the hierarchy appears to be maintained chiefly by subordinates and by low- rather than high-ranking animals. A similar conclusion was reached by Gartlan (1966) for captive groups of *C. aethiops*. This finding is important for the doubt that it casts on the existence of a distinct dominance drive; indeed, the motivational state of the dominant animal is irrelevant in the interaction situation. Rowell considers that in view of the high proportion of "misin- terpreted" approaches made by high-ranking baboons, in many cases the dominant animal fails to act "appropriately" with respect to relative status.

Gartlan showed that in the normal social life of a wild *C. aethiops* popula- tion, adult males were especially likely to be attacked, the aggressors being adult females. Adult males were the victims in 46 recorded instances and adult females in only 17.

One of the main functions of a hierarchy has generally been considered the reduction of aggression (c.f. Scott [1962]). However, it seems to be a general rule in primates that hierarchies are both more pronounced and more rigid under captive conditions, and that correlated with this are levels of aggression much higher than normally found in wild populations. It is diffi- cult to reconcile the previous data with the proposed function of hierarchy in reducing aggression.

Hall (1965) stated that the spread of individuals of a wild patas group during foraging, measured along and at right angles to the direction of move- ment was often of the order of 300-500 m. He also indicated that some animals might be foraging as far as 800 m from the remainder of the group. In contrast, the room in which the captive group was kept at Bristol measures 14 ft 6 in X 20 ft 10 in X 10 ft (Hall, Boelkins and Goswell [1965]). Thus animals which normally spend a good part of the day up to several hundred metres away from other group members are here confined to a room of which the longest axis is less than seven metres. In addition to the stress inevitably caused in this manner and expressed, as has been noted previously, by an increase in aggression, certain other behaviour patterns are also altered. Ter- restrial primate species such as baboons and patas monkeys spend the greater part of the day moving slowly in the home range, foraging and feeding. Indeed, most other behaviour can be considered as taking place against this background. In captivity this basic daytime pattern is replaced by one or two brief feeding periods which occupy at most an hour or so. This leaves up to eleven hours or more of daylight which would normally be taken up foraging, unoccupied and with animals abnormally crowded. It is to be expected that behavioural changes will occur.

Some evidence of the severity of changes occurring in captivity is provided in many studies. Zuckerman (1932) gave a list of the common diseases causing death in his colony of *Papio hamadryas;* these included valvular disease of the heart, atheroma, pericarditis, pneumonia, pulmonary abscess, pleurisy and empyema, colitis — often ulcerative — enteritis, pancreatitis, peritonitis and nephritis. Many of these diseases are known to be exacerbated by stress and some are classifiable as more or less pure examples of stress disease. Zuckerman also noted the high number of deaths from "sexual fights" which is indicative of a severely pathological social condition. Carpenter (1942) gives details of the high mortality rate of rhesus monkeys imported onto Cayo Santiago Island to live in conditions supposedly similar to the natural habitat. Chance (1956) gives details of the high incidence of disease in the colony of rhesus of which he studied the social structure. He also noted that at the beginning of the study the colony was under stress from a variety of non-social factors including the after-effects of the journey from India, a high incidence of dysentery and infection from tuberculosis. These three studies, it may be noted, were instrumental in establishing social dominance as a concept applicable to normal social structure.

Studies of crowding and consequent endocrine and social pathology have been carried out in several mammalian species, but not extensively as yet in primates. The most exhaustive studies so far reported in this area have been those of Christian (1961;1963), Barnett (1964) and Bronson and Eleftheriou (1964; 1965a; 1965b). Christian (1961) found that there was a positive correlation between position in the social dominance hierarchy and adrenal weight. Dominant mice had the lowest adrenal weights and showed greatest somatic growth, with the reverse occurring in subordinate animals. The adrenal weights of socially intermediate mice were between the two extremes.

Hamburg (1962) noted that conditioned fear and avoidance in *Macaca mulatta* produced increases in plasma hydrocortisone levels which were comparable to those observed following injections of large amounts of adrenocorticotrophic hormone. Hamburg also cites evidence indicating that epinephrine, norepinephrine and aldosterone are also secreted during emotional distress.

The function of these secretions is apparently to mobilise the availability of carbohydrate and fat as a preparation for the oxidative processes involved in muscular exertion. In other words, they prepare the animal for both rapid and sustained movement away from the stressful stimulation. Hamburg makes the further point that such a capacity for anticipatory mobilisation may well have had selective advantage in mammalian evolution, and that the trend towards increasing anticipatory powers, including the anticipatory regulation of physiological processes may well have been one of the crucial features of primate evolution.

The increase in social interaction and particularly in aggression under crowded captive conditions has already been noted. The particular conditions of captivity also ensure that escape and withdrawal from noxious stimuli is generally impossible. The evolution of an efficient anticipatory mechanism, associated with a highly developed nervous system would mean that abnormally high levels of hydrocortisone and adrenalin may well be chronic features of captive populations. In this connection, Mason and Brady (1964) have demonstrated the extreme sensitivity of the pituitary-adrenal cortical system in rhesus monkeys to environmental influences, as measured by plasma 17-hydroxycorticosteroids. Such conditions may be expected to lead eventually to degenerative changes. The dominance hierarchy may therefore be an example of a potentially adaptive response becoming nonadaptive where conditions prevent it being successful; Anderson (1961) has shown that in the house mouse, emigration is the first behavioural response to crowding. Initial interactions in the captive situation may therefore be of great significance in determining subsequent dyadic behaviour, which may not be based in any way on relative differences of biological fitness. The secretion of several substances, including serotinin, epinephrine and ACTH, which is affected by social factors, is also known to have effects on the learning process. Levine and Jones (1965) noted the enhancement of avoidance learning following injection of ACTH. Bronson and Eleftheriou (1965a) have shown that the adrenal responses (unbound corticosteroids) of mice placed in the presence of a trained fighter were much greater if subjects had previously experienced defeat. These workers (1965b) also demonstrated that the presence of a trained fighter mouse, without actual physical contact, was sufficient to produce adrenal enlargement in a mouse that had previously been physically defeated. The nature and direction of dyadic interactions may therefore be reinforced by physical and somatic changes and to that extent be irreversible. The dominance hierarchy may not therefore be a means of reducing aggression within the context of normal group structure, as much as a social artefact, symptomatic of social stress.

The data quoted so far in this section mainly demonstrate the type and extent of social change in captivity. There is, however, another main source of information on the variability of primate social systems of particular value in elucidating the precise ecological pressures moulding the society and therefore of defining the limits of variation in normal social structure. This is the comparison of single species in different habitat areas. Exhaustive studies of this nature are few as yet, but results are suggestive.

Jay ([in press], pers. comm.) in a comprehensive study of the distribution, ecology and behaviour of wild *Macaca mulatta* noted differences in the size of groups, sex ratios and behaviour according to the nature of the habitat in which they were found. Thus she was able to distinguish between forest, roadside and city rhesus which showed differences in social structure and behaviour, particularly an increasing gradient of aggression from forest to city. Southwick, Beg and Siddiqi (1965) also noted fewer wounded individuals in rural habitats and forest areas compared with urban areas.

Jay (1965) also studied the common Indian langur, *Presbytis entellus* in Orcha and Kaukori, and her description of the social structure of the species (stable, multi-male groups) contrasts sharply with the descriptions by Sugiyama (1965;1967) who, in the Dharwar area of Mysore recorded unstable units with one-male groups and all-male parties, with much higher levels of aggression.

Similarly, Gartlan and Brain (in press) report differences in the sex ratio of *C. aethiops* groups in different parts of the distribution range, with a tendency towards fewer adult males in areas with a severe dry season. Gartlan (1966) also reported differences in the frequency of appearance of various social patterns in two populations of *C. aethiops* and correlated these with ecological factors.

The results of these studies indicate that social structure in many widespread primate species is largely habitat — rather than species — specific. The implications of this finding are considerable for theories of primate social evolution, and indeed for the whole concept of primate social structure.

HABITAT, ROLE STRUCTURE AND SOCIETY

Wynne-Edwards (1962) defines society as an organisation capable of providing conventional competition. Whilst this is undoubtedly one characteristic, to consider society only in terms of a population regulatory mechanism — as is the tendency when using the concept of social dominance in field studies — represents a gross oversimplification, especially in the persistent groupings of primates. A basic characteristic of society is that there is adaptive differentiation of function between group members. At one level this may take the form only of temporal differentiation; one animal may feed whilst another maintains vigilance, or in birds, one may brood the clutch whilst the other feeds. Such roles, which do occur in primate societies are necessarily temporary and subject to reversals and alternation. At the other extreme there may be morphological or physiological differentiation for social roles, which again occurs to some extent in primate social structure, but which is most pronounced in the social insects.

In the same way in which morphological and physiological adaptations reflect accommodation to particular selective pressures, social systems are appropriate to particular environmental conditions. It is, however, both possible and necessary to distinguish between adaptations at different evolutionary levels. Thus mammals, as a class, have certain characteristics in common and may therefore be expected to show certain similarities. Further down the scale, one expects to find a given species only within certain limits of habitat variation. Comparative studies are therefore essential before it may be stated with any degree of confidence whether a particular pattern of social organisation is characteristic of only a particular social group, or of the population, species or genus. In the past, precise analysis of social structure has been difficult due to the lack of a consistent descriptive system capable of dealing effectively with the complexity of social systems in the wild. However, one scheme which provides at least some possibility of quantitative comparison is the identification of social roles of adaptive importance and the construction of social role profiles for groups and individuals.

Previous attempts have been made to describe primate social systems in terms of social roles. Bernstein and Sharpe (1966) described some of the roles important in a captive group of rhesus monkeys. One of the characteristic roles of the dominant male was concerned with "punishing" or interfering in aggressive episodes between other group members before aggression became socially disruptive. A similar role of the dominant male of a captive group of *Papio anubis* was noted by Rowell (1966). A second important role of the dominant male in the captive rhesus group was in acting as a protective buffer between the group and potential predators or aggressors. The dominant male was also a "focus and influence" of adult female activity. The main roles of adult females were in checking the aggressiveness of adult males and in acting as "social foci." The role of the low-ranking animals, which were more or less peripheral and solitary and generally alert, was possibly in warning the group of potential danger, although in captivity it seems likely that the main focus of attention would be towards intragroup social interactions rather than extra-group events.

Gartlan (1966) described a complexly differentiated social structure in a wild population of *C. aethiops* on Lolui Island, Lake Victoria, Uganda. The habitat comprised grassland with stands of moist semi-deciduous rain forest and the typical territory of a monkey group consisted of a forested "core area" from which expeditions into grassland were made. Some of the results of that study will now be described, since they indicate the extent to which social roles are adaptive to particular environmental pressures. This analysis will only be concerned with roles of adaptive significance for this particular habitat; play and maternal behaviour are therefore excluded, not because they are of no importance socially, but because their main effects are long-term and outside the scope of this particular analysis. It should also be pointed out that social roles are not necessarily active, although in practice this is generally so; animals receiving a large number of friendly approaches have a distinct social role. The social roles described here involve only 65 observation hours of a single group which included at the time 20 individuals, 3 adult males, 6 adult females, 1 juvenile male, 4 subadult females (physically mature but nulliparous) and 6 infant-twos. Of the three adult males, it should be noted, only one, A♂1 was with the group throughout the entire study. A♂2 had joined the group in August 1963, nine months prior to the sample period, and A♂3 joined in November of the same year.

The social interactions observed in the sample period were classified in discrete categories.

1. Territorial display: jumping around (Ja).
2. Social vigilance: looking out (Lo) .
3. Social focus: the object of friendly approaches; not including the behaviour of mothers to their own infants (Sf).
4. Friendly approach (Fa).
5. Territorial chasing: the chasing out and exclusion of intruders from the group's territory (Tc).
6. Punishing: interfering in intra-group aggression; only one incident recorded in this sample (Pu).
7. Leading: the initiation of compact group movements across grass-land, fairly rare in this group's wooded territory (Le).

The complete record of all social interactions, excluding maternal behaviour and play, which occurred in the sample period, could be described using these seven categories. All incidents in each role category were then summed and the percentage contribution of each animal assessed. It is thus possible to construct a matrix between the role categories and the contributions of either individuals or age-sex classes. The matrix for the latter is shown in Table I. Statistical analyses of such matrices may be used to determine the significance of social role patterns within groups.

Expected frequencies for the contribution of each age-sex class may be calculated from the known number of individuals in the group. These were, adult males 15%, adult females 30%, juvenile male 5%, subadult females 20% and infant-twos 30%. A chi-square test for each role category determines whether the percentage contribution of each age-sex class is other than would be expected by chance. Calculations indicate that the distributions shown for each social role are all significant at the 0.001 level of confidence, except for Sf which is significant at the 0.01 level. However, the importance of learning in the aetiology of social roles is known from several studies (e.g., Kawai [1958]) and may be expected to apply especially to those roles which are specific to particular habitats. It is possible then to argue that to include figures for infant-twos in statistical analyses may introduce a bias from the inclusion of a relatively large subgroup with relatively few social roles. The percentages shown in Table I may therefore be recalculated excluding this age class and with a recalculation of the expected frequencies. If this is done, all role distributions are still significant at the 0.001 level, except for Sf which this time is not significant. It may be concluded therefore that with the possible exception of "social focus," social roles are not randomly distributed throughout the population, but show differentiation according to age-sex class. Graphical illustration of the differentiation in social roles between adult males and adult females is shown in Figure 1.

Differentiation of social roles also occurs within, as well as between, age-sex classes. Illustrated in Figure 2 are the social roles for the adult males in the group. It is clear that several of the distinctive male roles, such as jumping around and territorial chasing, are here performed only by one adult male, A♂1. A♂1 was also the only one of the three observed to make friendly

TABLE I.
Percentage contribution by age-sex class to group role structure.

| Social role category | Age-sex class | | | | |
	A♂	A♀	J♂	S♀	I2
Ja	66%	–	33%	–	–
Lo	35%	38%	3%	12%	12%
Sf	12%	46%	4%	27%	12%
Fa	3%	32%	–	47%	15%
Tc	66%	–	33%	–	–
Pu	100%	–	–	–	–
Le	32%	49%	–	16%	–

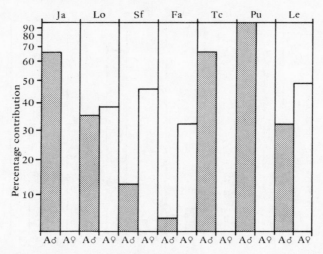

Fig. 1. Adult male and adult female social roles. Total A♂s = 3. Total A♀s = 6. Ja = Jumping around, Lo = Looking out, Sf = Social focus, Fa = Friendly approach, Tc = Territorial chasing, Pu = Punishing, Le = Leading.

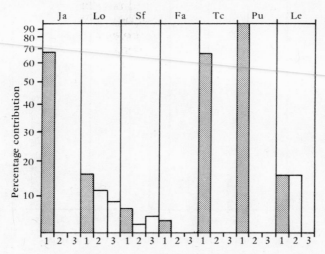

Fig. 2. Individual adult male social roles. 1 = A♂1, 2 = A♂2, 3 = A♂3.

approaches to other group members. The only role in which joining males scored highly was in social vigilance, although A♂2 was observed to initiate compact group movements in the same frequency as A♂1.

Other social role profiles for individual animals are shown in Figures 3 and 4. Figure 4 is a comparison of the roles of an adult male (A♂1) and the juvenile male (J♂) of the group. It is clear that the juvenile animal resembles

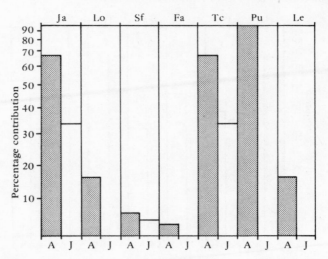

Fig. 3. Adult Male and Juvenile Male Social Roles. A = A♂1, J = J♂.

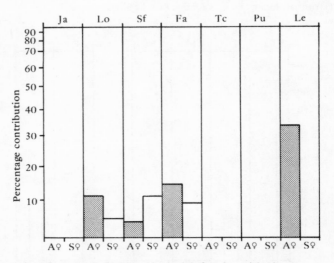

Fig. 4. Adult female and subadult female social roles.

this adult male more closely in the distribution of roles than do other adult males of the group. J♂ was observed to perform both territorial jumping-around and territorial chasing, although not as frequently as A♂1. It was not observed to perform intra-group punishing. Figure 4 indicates social roles for an adult female (A♀) and a subadult female (S♀). The relatively high score for the friendly approach roles in both of these age-sex classes is noteworthy, especially in comparison with the low scores for the males. Subadult females did not initiate compact group movements as frequently as did adults.

These data suggest that there are both maturational and social factors in the assumption of social roles; juveniles performing the same roles as adults do not do so as frequently. But perhaps more important is the indication that in adult males the full performance of the typically adult male roles depends to a large extent on the length of time the male in question has been in the group, and presumably therefore on social factors such as "confidence." In other groups on the island two or three adult males were observed to perform the typically adult male roles. The adult males in the sample group joining during the course of the study, whilst presumably capable of performing typically adult male social roles, were thus probably inhibited by social factors from doing so.

The analysis of primate social structure in terms of social role differentiation permits the identification of environmental pressures important in moulding the society. Among pressures known to be important, as indicated in the preceding analysis, are population density, the type and availability of food resources, and predation pressures. Analysis of this type follows logically from the proposition that social structure is determined multifactorially and is appropriate to particular ecological conditions. Comparative studies, experimental alteration of the habitat and developmental studies will also indicate which social roles and patterns are species-specific and which are adaptive to particular environmental conditions. The identification of social roles thus permits comparison within and between species both objectively and quantitatively. From such studies, predictions about social structure in particular habitat conditions become possible. This is not the case with social dominance theory which has been used both implicitly and explicitly in theories of primate social evolution and has necessarily influenced both the nature and direction of evolutionary theories. It is proposed here to examine certain of these conclusions.

TRENDS APPARENT IN PRIMATE SOCIAL EVOLUTION

Attempts at identifying the main trends apparent in primate social evolution have until recently been hampered not only by concepts derived from social dominance theory, but also by a lack of information about the general biology and social structure of arboreal Cercopithecid primates. This gap is now beginning to be filled by very recent and continuing studies such as those of Aldrich-Blake (pers. comm.) on *Cercopithecus mitis*, Chalmers (pers. comm.) on *Cercocebus albigena*, Gartlan on *Cercopithecus nictitans martini*, *C. mona* and *C. erythrotis camerunensis*, and Struhsaker (pers. comm.) on several rain-forest species in West Africa. With these new data a more comprehensive attempt at an evolutionary scheme is permitted.

In a recent paper, Crook and Gartlan (1966) devised a general classificatory scheme in which the various social systems described up to that time were arranged into five adaptive grades (following Huxley [1959]). These grades were dependent on such factors as habitat, diet, size of groups, reproductive units, sexual dimorphism and population dispersion. One of the main conclusions to be drawn from this scheme is that there is no evidence of simple, linear social evolution from the arboreal, nocturnal Lemuriformes to the anthropoid apes. Grade III, for example, included *Lemur macaco, Alouatta palliata* and *Gorilla,* indicating parallel development of social structure in these diverse species. This scheme was not designed to deal with the observed intra-specific variability which has been widely reported, but was concerned with species-specific social structure.

A paper by Frisch (in press) attempts to deal with both species-specific social structure and intra-specific variability in a theory of primate social evolution. The two species considered by Frisch, and from which he draws his evolutionary conclusions, are *Macaca fuscata* and *Presbytis entellus.* Both have been extensively studied by Japanese workers. The typical social group of *M. fuscata* has a characteristic spatial distribution, with animals of particular age-sex classes and statuses characteristically located in particular zones relative to the rest of the group. There are relatively high levels of aggression, and groups are characterised by "inventive" behaviour (Tsumori [1967]) and "precultures" (Itani [1958]). These two terms indicate generally the location of new food sources by a behaviour pattern new to the repertoire of the group, and the mode and pattern of assimilation by the group as a whole. This typical social structure occurs, with only minor variations, in all populations and habitats studied. *P. entellus,* on the other hand, shows very different social structure in different habitat areas. All-male parties, one-male bisexual groups and multi-male bisexual groups occur in different environments. There is no evidence of "inventive" behaviour or of "precultures" and no characteristic pattern of spatial distribution.

Frisch distinguishes between variability in the basic structure of groups *(P. entellus)* and variability in individual behaviour with a stable basic structure *(M. fuscata).* In the former case, Frisch suggests, langur social structure may reflect a predominantly passive adaptation to environmental pressure, evolved by direct selection, with the habitat determining absolutely the form of the social structure. In the case of the macaque, it is suggested that this species, with an invariant social system, creates its adaptations by inventing ways of exploiting the opportunities offered by the environment — the development of "precultures." The macaque, in other words, through the development of individually variable behaviour, is capable of a limited degree of control over the environment.

Frisch notes that fossil jaws and teeth showing characteristics of the Colobinae have been found in Miocene deposits of East Africa, whilst dentitions clearly belonging to the macaque-baboon group begin to appear only in the Upper Pliocene and Lower Pleistocene. The implication is that the langur has preserved a rather primitive type of social structure compared with the macaque-baboons. The more structured society, characterised by more social dominance interactions is, in other words, considered to be a more recent evolutionary development.

In my view this analysis arises from an oversimplification due to the different origins of the data. Those of *M. fuscata* for example have, with few exceptions, been derived from populations which have been artificially fed or "provisionised," food being scattered in a restricted area into which the group must come, so causing an increase in crowding. The langur studies, in contrast, were carried out without artificial feeding of any kind. The aim of artificial feeding is the study of groups which are, for a variety of reasons, difficult to observe under normal conditions. However, the technique has results not dissimilar from those of captivity; day-ranging patterns are altered, the spatial distribution of the group in the territory is much reduced and social interaction, particularly aggression, increases. Gartlan (1966) in an experiment on artificial feeding of a wild group of C. *aethiops* noted an increase in aggression from a predetermined, natural basal level of 0.28 incidents an hour to one every 29.5 min — more than a sevenfold increase. The alterations to be expected under these circumstances are therefore likely to be different only in degree from those obtained in captivity.

The uniformity of M. *fuscata* social structure may therefore be an artefact of the provisionising technique. Jay's work ([in press]; pers. comm.) on the variability in group size, composition and social structure in M. *mulatta* indicates that the typical social structure described for M. *fuscata* does not occur invariably throughout the *Macaca* genus, signifying the uncertain status of phylogenetic explanations of the observed social differences. There is also no evidence that the differences between M. *fuscata* and P. *entellus* social structure represent relatively more advanced or more primitive social systems. This is highlighted by comparative evidence available from recent studies of lemurs (especially that of Jolly [1967]). The social systems of this family show considerable variation and are generally unlike the one-male and all-male groups described by Sugiyama (1965; 1967) in P. *entellus*. The social structure of *Lemur catta* is appropriate to the particular environmental conditions characteristic of the species. Under similar habitat conditions it is not unlike the social structure of African catarrhine monkeys. The use of the term "primitive" in discussions of social structure clearly requires adequate justification and evidence that differences are not phenotypic in origin.

In spite of the possible artefactual nature of differences in social structure so far described between P. *entellus* and M. *fuscata*, social patterns such as the transmission of "precultures" and "inventive" behaviour in M. *fuscata* clearly require an explanation as neither of these patterns has been recorded in P. *entellus*.

It is clear from Japanese accounts of precultural transmission in M. *fuscata* that new behaviour patterns generally originated in infant-twos. Highly developed curiosity is characteristic of many young mammals and particularly of primates, so that one of the main patterns of maternal behaviour is in preventing or rescuing infants from potentially dangerous situations into which they threaten to stray or have actually strayed. Gartlan (1966) noted pronounced curiosity in infant-twos of C. *aethiops*. In the presence of the human observer an entire play-group would frequently come and sit in a tree under which the observer was seated, gradually approaching closer and closer, peering intently at the observer and showing neither threat nor alarm unless he moved suddenly. A highly characteristic posture was frequently observed during this behaviour. The infant, from a quadrupedal position on a branch fairly near the observer would gradually and slowly invert itself until it was completely upside down, all the time looking intently at the observer. This was interpreted as an attempt to vary the stimulus qualities of the object. The behaviour disappeared as soon as play-groups disappeared; no examples of it were ever observed in older animals.

It is possible, therefore, that savanna and urban habitats might merely provide ecologically appropriate scope for the inquisitiveness of infants; some of the investigations will lead to practical results which may be passed on to other group members. In the forest, behaviour of this type may occur, but the necessity for it will be less; the diet of forest species is generally restricted to fruit and leaves, with the latter widely available throughout the year. *P. entellus* has the specialised stomach characteristic of the Colobinae, functional in a diet with a high proportion of mature leaves. On the savanna, on the other hand, a greater variety of food items is usual, ranging from small mammals, insects and arachnids, bulbs, fungus, leaves to stems and fruits. The savanna provides more chance for the new behaviour patterns to be of biological value. This does not mean that forest species are incapable of this type of behaviour, but merely that it is inappropriate. The habitat determines population size and structure of groups and has secondary effects on both social interactions and the complexity of individual behaviour. Social patterns such as "inventive" behaviour and "precultures" shown in species such as *M. fuscata* may therefore be phenotypic and not associated with a higher level of social evolution than is characteristic of forest species.

DISCUSSION

Field studies of primate social organisation and behaviour have generally utilised the concept of social dominance as a structuring mechanism. This concept, along with that of the sexual bond, was derived largely from early studies of captive groups. Recent studies of avian social organisation have demonstrated multifactorial determination of social structure, a discovery which almost certainly holds true for primate groups and renders impractical the analysis of primate social structure in terms of variation in strength of a unitary structuring mechanism.

Social dominance theory makes several implicit assumptions, including (1) that a continuum of rank-order criteria exists throughout groups, and (2) that dominance is a cluster of inter-related behaviour patterns. The former has been demonstrated not to occur, and in the latter case behaviour patterns often used as indices of social dominance often show no inter-correlation. There is, in addition, the problem of why intense selection for the associated morphological and behavioural phenomena does not occur if sexual behaviour is dependent on dominance characteristics. Evidence indicates that, on the contrary, learning plays a significant part in the assumption of social roles, and that genetic influences are minimal.

The social group was originally considered as a "harem" organised and maintained by the adult male, who was considered to function in controlling both the structure and the behaviour of the group. Ecological studies have since demonstrated the inadequacy of this concept of social structure. The group is an adaptive unit, the actual form of which is determined by ecological pressures. Different roles of relevance to particular ecological conditions are performed by different animals. There is no a priori reason why either more aggressive or larger animals should be any more efficient than their peers at performing social tasks such as the leading of the group or the detection of predators, and in fact they are not observed to be more efficient in these roles. Biological equipment however determines to some extent the range of roles which an animal performs. There is no evidence that particular roles are associated with different levels of biological fitness; the performance of particular roles seems to depend on maturation, learning and social circumstances. Changes in circumstances result in different animals performing roles. If the dominant male is removed from a captive group, another animal will take over the roles. If the original dominant is then replaced, he is often defeated in the ensuing fight.

There remains the problem of why social dominance came to be accepted as the normal structuring mechanism of primate societies. There is a great deal of evidence indicating that the early studies which contributed so much towards the establishment of the concept as a normal structuring mechanism, were in fact studies of populations under severe social stress.

There is evidence that the primates have evolved a highly efficient system for the anticipatory mobilisation of carbohydrate and fat, permitting both rapid and sustained movement away from the group in conditions of social stress. Under crowded conditions in the wild, emigration would be possible before actual fighting took place, which may well have had selective advantage and be one possible mechanism of adaptive radiation. Captivity approximates in many ways to conditions of severe crowding and there is evidence that under these conditions primates are subject to chronic physiological involvement. The particular conditions of captivity, whilst increasing aggressive social interaction, do not permit escape from the stressing conditions, and thus lead to degenerative physiological changes. There is evidence from other studies of mammalian populations that the dominance hierarchy is correlated with changes in the weight of the adrenal body. Such changes result from aggressive encounters, from encounters without aggression after initial defeat, and even from mere physical proximity to aggression. The result of an initial encounter between animals in the captive situation may therefore be crucial in determining subsequent dyadic behaviour, and such a result may depend on many different factors. An animal living in a group of which it has a lifetime's experience will clearly have an advantage in any aggressive episode with a newcomer. Thus confidence, and a variety of social factors may well be the determining factor in the results of dyadic interactions and therefore in the formation of a social hierarchy. The original social structure of the group may only be of importance inasmuch as it will inevitably affect the confidence of animals involved in social interactions, and therefore indirectly affect the results of dyadic interactions in the captive situation.

The function of a hierarchy has often been considered as the reduction of aggression. Yet field studies of species that have previously been studied only in captivity show without exception that in these latter circumstances hierarchies are both more pronounced and more rigid, and that aggression is much more common. It is difficult to reconcile these data with the proposed function of reducing aggression.

Social dominance has been used in theories of primate social evolution. It has been thought that more structured groups showing more dominance interactions and other behaviour patterns such as "inventive" behaviour and "precultures" are more advanced than those with a more variable basic social structure and which fail to show "precultures." However, the data on the more structured groups are data obtained from "provisionised" groups, and artificial feeding has results different only in degree from those of captivity. The stable basic structure may therefore be an approximation to the conditions observed in captivity. Comparative data from recent studies of lemur social structure provide a potential model of primitive social organisation. In fact, in this case the social structure again seems to be habitat-specific. The labels "advanced" or "primitive" clearly require extensive justification in theories of primate social evolution.

However, there remains the problem of "precultures" and "inventive" behaviour, which have been so carefully documented by Japanese workers. The inquisitiveness of infants is well-documented and it seems likely that in savannas or urban habitats this behaviour may well have results of biological value. In fact, it is in this age-class that the precultures originate. It is clear that the type of habitat will limit the potentiality of such responses, and their assimilation into the group repertoire. There is no evidence that species not showing "precultures" are incapable of their development and therefore less advanced socially than those species showing such behaviour patterns. It is a question of present relevance, not of past evolution.

Summary

The origins of the concept of social dominance are described. The implications and assumptions made by the theory are listed and evaluated and the value of the concept to studies of free-ranging primates is assessed. Possible functions of social dominance hierarchies are considered and evaluated. Studies in the physiological and endocrine responses to crowding and stress are cited and their relevance to the status of social dominance theory indicated. The shortcomings of social dominance theory as both a descriptive and an analytical concept are described and an alternative scheme put forward — the description of social structure in terms of social role differentiation. The effect of social dominance on theories of primate social evolution is assessed, and an alternative explanation of observed differences suggested.

ACKNOWLEDGMENTS

The studies reported here of *C. aethiops* under the author's name were undertaken whilst he was in receipt of a Medical Research Council Scholarship for Training in Research Methods. Support from the Wenner-Gren Foundation for Anthropological Research is also gratefully acknowledged.

The author wishes to thank Dr. J.H. Crook, Mr. J. Archer and Mr. P. Aldrich-Blake for critical appraisal of this manuscript. Special thanks are also due to Dr. J.D. Goss-Custard for extensive and valuable discussions.

REFERENCES

Allee, W.C. 1952. Dominance and hierarchy in societies of vertebrates. In Structure et Physiologie des Sociétés Animales, pp. 157-81. Paris: Centre de la Recherche Scientifique.

Anderson, P.K. 1961. Density, social structure and non-social environment in house mouse populations and the implications for regulation of numbers. Trans. N.Y. Acad. Sci. (II)23:447-51.

Barnett, S.A. 1964. Social stress. Viewpoints in Biology 3:170-218.

Bernstein, I.S., and Sharpe, L.G. 1966. Social roles in a rhesus monkey group. Behaviour 36:91-104.

Brian, A.D. 1949. Dominance in the great tit, *Parus major.* Scot. Nat. 61:144-55.

Bronson, F.H., and Eleftheriou, B.E. 1954. Chronic physiological effects of fighting in mice. Gener. comp. Endocr. 4:9-14.

_____ . 1965a. Adrenal response to fighting in mice: separation of physical and physiological causes. Science 147:627-28.

_____ . 1965b. Behavior, pituitary and adrenal correlates of controlled fighting in mice. Physiol. Zool. 38:406-411.

Carpenter, C.R. 1942. Sexual behavior of free-ranging rhesus monkeys, *Macaca mulatta.* I.J. comp. Psychol. 33:113-42.

Chance, M.R.A. 1956. Social structure of a colony of *Macaca mulatta.* Brit. J. Anim. Behav. 4:1-13.

Christian, J.J. 1961. Phenomena associated with population density. Proc. nat. Acad. Sci., Wash. 47:428-49.

_____ . 1963. The pathology of overpopulation. Milit. Med. 128(7):571-603.

Crook, J.H. 1965. The adaptive significance of avian social organisation. Symp. zool. Soc. Lond. 14:181-218.

Crook, J.H., and Gartlan, J.S. 1966. The evolution of primate societies. Nature, Lond. 210:1200-1203.

DeVore, I. 1965. Male dominance and mating behavior in baboons. In Sex and Behavior, ed. F. Beach, pp.266-89. New York: John Wiley & Sons.

Frisch, J.E. Variability in primate social behavior: its evolutionary significance. In Explorations in Primate Behavior, ed. P. Jay. New York: Holt, Rinehart & Winston, in press.

Gartlan, J.S. 1964. Dominance in east African monkeys. Proc. E. Afr. Acad. 2:75-79.

_____ . 1966. Ecology and behaviour of the vervet monkey, *Cercopithecus aethiops pygerythrus,* Lolui Island, Lake Victoria, Uganda. Ph.D. Thesis, University of Bristol.

Gartlan, J.S., and Brain, C.K. Ecology and social variability in *Cercopithecus aethiops* and *C. mitis.* In Explorations in Primate Behavior, ed. P. Jay. New York: Holt, Rinehart & Winston, in press.

Hall, K.R.L. 1962. The sexual, agonistic and derived social behaviour of the wild Chacma baboon, *Papio ursinus.* Proc. zool. Soc. Lond. 139:283-327.

_____ . 1965. Behaviour and ecology of the wild patas monkey, *Erythrocebus patas,* in Uganda. J. Zool. Lond. 148:15-87.

Hall, K.R.L.; Boelkins, R.C.; and Goswell, M.J. 1965. Behaviour of patas monkeys, *Erytbrocebus patas*, in captivity, with notes on the natural habitat. Folia primat. 3:22-49.

Hall, K.R.L., and DeVore, I. 1965. Baboon social behavior. In Primate Behavior: Field Studies of Monkeys and Apes, ed. I. DeVore, pp. 53-110. New York: Holt, Rinehart & Winston.

Hamburg, D.A. 1962. Relevance of recent evolutionary changes to human stress biology. In Social Life of Early Man, ed. S.L. Washburn, pp. 278-88. London: Methuen & Co.

Hazama, N. 1964. Weighing wild Japanese monkeys in Arashiyama. Primates 5(3-4):81-104.

Huxley, J.S. 1959. Clades and grades. In Function and Taxonomic Importance. Systematics Assn. Publn. 3:21-22.

Itani, J. 1958. On the acquisition and propagation of a new food habit in the troop of Japanese monkeys at Takasakiyama. Primates 1:84-96.

Jay, P. 1965. The common langur of North India. In Primate Behavior. Field Studies of Monkeys and Apes, ed. I. DeVore, pp. 197-249. New York: Holt, Rinehart & Winston. Also in Explorations in Primate Behavior, ed. P. Jay. New York: Holt, Rinehart & Winston, in press.

Jolly, A. 1967. Lemur Behavior, Chicago: University of Chicago Press.

Kawai, M. 1958. On the system of social ranks in a natural troop of Japanese monkeys. I. Basic and dependent rank. Primates 1-2:111-30. Translated in S.A. Altmann, Japanese Monkeys, pp. 66-86. Edmonton: University of Alberta Press, 1965.

Kawamura, S. 1958. Matriarchal social ranks in the Minoo-B troop: a study of the rank system of Japanese monkeys. Primates 1-2:149-56. Translated in S.A. Altmann, Japanese Monkeys, pp. 105-112. Edmonton: University of Alberta Press, 1965.

Köhler, W. 1925. The Mentality of Apes. New York: Harcourt.

Kummer, H. 1957. Soziales Verhalten einer Mantelpavian-Gruppe. Beiheft Schweiz. Z. Psychol. 33:1-91.

Lack, D. 1966. Population Studies of Birds. Oxford: Clarendon Press.

Lancaster, J.B., and Lee, R.B. 1965. The annual reproductive cycle in monkeys and apes. In Primate Behavior. Field Studies of Monkeys and Apes, ed. I. DeVore, pp. 486-511. New York: Holt, Rinehart & Winston.

Levine, S., and Jones, L.E. 1965. Adrenocorticotrophic hormone (ACTH) and passive avoidance learning. J. comp. physiol. Psychol. 59:357-60.

Maslow, A.H. 1936. The role of dominance in the social and sexual behavior of infrahuman primates. I. Observations at Vilas Park Zoo. J. genet. Psychol. 48:261-77.

Maslow, A.H., and Flanzbaum, S. 1936. The role of dominance in the social and sexual behavior of infra-human primates. II. An experimental determination of the behavior syndrome of dominance. J. genet. Psychol. 48:278-309.

Maslow, A.H.; Rand, H.; and Newman, S. 1960. Some parallels between sexual and dominance behavior of infra-human primates and the fantasies of patients in psychotherapy. J. nerv. ment. Dis. 131:202-212.

Mason, J.W., and Brady, J.V. 1964. The sensitivity of psychendocrine systems to social and physical environment. In Psychobiological Approaches to Social Behavior, eds. P.H. Leiderman and D. Shapiro, pp. 4-23. Stanford, Calif.: Stanford University Press.

Mason, W.A. 1960. Socially mediated reduction in emotional responses of young rhesus monkeys. J. abnorm. soc. Psychol. 60:100-104.

McDougall, W. 1908. Social Psychology: An Introduction. London: Methuen & Co.

Murchison, C. 1935. The experimental measurement of a social hierarchy in *Gallus domesticus*. III. The direct and inferential measurement of social reflex No. 3. J. genet. Psychol. 46:76-102.

Rowell, T.E. 1966. Hierarchy in the organization of a captive baboon group. Anim. Behav. 14:430-443.

_____. 1967. Female reproductive cycles and the behavior of baboons and rhesus macaques. In Social Communication Among Primates, ed. S.A. Altmann, pp. 15-32. Chicago: University of Chicago Press.

Sade, D.S. 1964. Seasonal cycle in size of testis in free-ranging *Macaca mulatta*. Folia primat. 2:171-80.

———. 1967. Determinants of dominance in a group of free-ranging rhesus monkeys. In Social Communication Among Primates, ed. S.A. Altmann, pp. 99-114. Chicago: University of Chicago Press.

Sade, D.S., and Hildreth, R.W. 1965. Notes on the Green Monkey, *Cercopithecus aethiops sabaeus* on St. Kitts, West Indies. Carib. J. Sci. 5(1-2):67-81.

Schaller, G.B. 1963. The Mountain Gorilla: Ecology and Behavior, p. 431. Chicago: University of Chicago Press.

Schjelderup-Ebbe, T. 1931. Die Despotie im sozialen Leben der Vögel. Völker-psychol. Sozialog. 10(2):77-140.

Scott, J.P. 1962. Hostility and aggression in animals. In Roots of Behavior, ed. E.L. Bliss, pp. 167-78. New York: Harper & Row.

Simonds, P.E. 1965. The bonnet macaque in South India. In Primate Behavior. Field Studies of Monkeys and Apes, ed. I. DeVore, pp. 175-96. New York: Holt, Rinehart & Winston.

Skard, A.G. 1937. Studies in the psychology of needs: observations and experiment on the sexual need in hens. Acta psychol., Amst. 2:172-232.

Southwick, C.H.; Beg, M.A.; and Siddiqi, M.R. 1965. Rhesus monkeys in North India. In Primate Behavior. Field Studies of Monkeys and Apes, ed. I. DeVore, pp. 111-59. New York: Holt, Rinehart & Winston.

Sugiyama, Y. 1965. On the social change of Hanuman langurs (*Presbytis entellus*) in their natural habitat. Primates 6(304):381-429.

———. 1967. Social organization of Hanuman langurs. In Social Communication Among Primates, ed. S.A. Altmann, pp. 221-36. Chicago: University of Chicago Press.

Taylor, J.C. 1966. Home range and agonistic behaviour in the grey squirrel. In Play, exploration and territory in mammals, eds. P.A. Jewell and C. Loizos. Symp. zool. Soc. Lond. 18:229-34.

Tsumori, A. 1967. Newly acquired behavior and social interactions of Japanese monkeys. In Social Communication Among Primates, ed. S.A. Altmann, pp. 207-19. Chicago: University of Chicago Press.

Washburn, S.L., and DeVore, I. 1961. The social life of baboons. Sci. Amer. 204:62-71.

Wynne-Edwards, V.C. 1962. Animal Dispersion in Relation to Social Behaviour. London and Edinburgh: Oliver & Boyd.

Yerkes, R.M., and Yerkes, A.W. 1929. The Great Apes: A Study of Anthropoid Life. New Haven: Yale University Press.

Zuckerman, S. 1932. The Social Life of Monkeys and Apes. London: Kegan Paul, Trench & Trubner.

Eight

EDITOR'S INTRODUCTION

William Mason discusses primate research as it relates to the solution of human problems and isolates three general research strategies. The first, which treats the nonhuman primate as a model of man, a biological near-equivalent, is of primary importance to medical research. The second strategy is one associated with the "building of a conceptual model of some behavioral mechanism or process," and it involves abstracting from the whole behavior of nonhuman primates "part-systems" which are then studied as if they were "self-contained functional entities." Mason, whose own laboratory research on determinants of social behavior in rhesus monkeys and chimpanzees has brilliantly exemplified this strategy, argues that if its results are to increase significantly our understanding of man it must be fitted into a broader, evolutionary-comparative approach. This approach, his third strategy, is of course the one which previous papers in this reader have represented. Mason goes on to present a provocative argument for behavior neoteny, the retention of juvenile (behavioral) traits into adulthood, as an important trend in primate evolution, but his paper is included in this selection primarily for its concise position statement on the relationship between the conceptual model approach and the evolutionary-comparative perspective.

Partly as a consequence of the bewildering diversity observable in the behavior of the nonhuman primates, there has developed over the past decade a tendency on the part of some primatologists to discourage the application of inferences drawn from the behavior of the nonhuman primates to the behavior — especially the social behavior — of man. Mason admits that the study of nonhuman primates is not the most straightforward approach to investigating problems of human behavior, but he holds that "in electing to experiment with monkeys or apes the behavioral scientist makes certain assumptions about the relationship between human behavior and that of the other primates" and that "these assumptions are the foundations of research strategy: they influence not only the kind of research that is performed, and the character of the results, but — since these findings provide raw materials for inferences about human needs and human actions — they contribute to our views on the nature of man." If primate research needed any justification, it seems to me that it would be provided in this statement. On the other hand, the papers collected in this reader are not direct research reports but have to do with research strategy and with specific theories of evolution derived from the data of research, and I hope that the converse of Mason's statement may have already occurred to the reader. The behavioral scientist *starts out* with certain views about the nature of man which do much to condition his assumptions about the relationship between human behavior and that of the other primates. It goes without saying that complete objectivity in the observation and interpretation of our nonhuman primate relatives is beyond the reach of the human observer.

Scope and Potential of Primate Research

BY WILLIAM H. MASON

INTRODUCTION

We are in a time of unprecedented boom in primate research. During the present decade an international primatologic society has been formed, several scientific journals have been established devoted exclusively to the primates, and the number of books and monographs on primate behavior has more than doubled over that of the preceding ten years. These are but the products of a rapidly expanding interest. Extrapolating from present trends, some observers predict that the scientists watching primates in the wilds of Africa will soon outnumber them. Laboratory research is showing comparable growth. Many of our institutions of higher learning have major primate research facilities, and many others harbor at least a modest colony of rhesus or squirrel monkeys in some obscure corner of the campus. Perhaps the clearest single indication of the magnitude of our national commitment, however, is the establishment by the Federal Government of several major centers throughout the United States to support and encourage research with the nonhuman primates.

The stated purpose of these centers, and the orientation of many of the smaller facilities as well, is to advance knowledge of the biologic characteristics of the nonhuman primates as it relates to the health of man. To look upon the monkeys and apes as a special resource for getting at human problems makes good sense, of course, since these animals are, after all, our closest living kin. Besides, there have been many examples within recent years in which their usefulness has been demonstrated: We are all familiar with their role in the early production of polio vaccines, with the part they played in our first ventures into space, and with the present interest in them as possible donors of replacement organs for human patients. Recognizing the biologic affinities between the nonhuman primates and man, and knowing what they have helped us to achieve thus far, most scientists are probably satisfied that whatever they are doing, if the work is being conducted on monkeys and apes and it is being done well, it will necessarily have a bearing on human problems.

From Jules H. Masserman, ed., *Science and Psychoanalysis*, Vol. XII, pp. 110-112. Reprinted by permission of Grune & Stratton, Publishers.

I must admit to sharing fully this belief that research with the nonhuman primates will, in the long run, help to improve the human condition. And, as a psychologist, I naturally assume that this faith is as appropriate for behavioral investigations as for other kinds of research.

At the same time, however, in view of the obvious psychologic differences between the nonhuman primates and man, I cannot help wondering what shape our behavioral contributions will take, and whether we are doing our best to make the most of the research opportunities that are available to us right now. I don't mean to imply that these questions are a special preoccupation of mine — like everyone else, I proceed with my research, entirely confident most of the time that I am on the right track — nor would I presume to air them here, except that they form the root of my concern with the problem of the scope and potential of primate research.

But achieving a balanced overview of this vast and heterogeneous field — to say nothing of suggesting its potential significance — is clearly beyond the compass of this chapter. I propose, instead, to limit myself to a discussion of some of the strategies of primate research as they relate to the solution of human problems. Although this statement might suggest the opening phrases of a discourse on method, I can assure you that research techniques and experimental design will not be my concern. My concern is the scope and potential of primate research; but I intend to approach this question by examining certain views and expectations that are likely to be held by a scientist using monkeys or apes as a means for investigating problems of human behavior. At first glance this must seem a devious route indeed for approaching the subject, but my thesis is actually quite simple and straight-forward: First, I hold that in electing to experiment with monkeys or apes, the behavioral scientist makes certain assumptions about the relationship between human behavior and that of the other primates — even though he may not take the time to formulate them precisely. Secondly, I believe that these assumptions are the foundation of research strategy: They influence not only the kind of research that is performed, and the character of the results, but — since these findings provide raw materials for inferences about human needs and human actions — they contribute to our views on the nature of man. And isn't this the ultimate hope that most of us cherish for primate research, that it will enlarge, and enrich, and improve our understanding of human nature? Isn't this the measure of the scope and the potential of primate research that most deeply concerns us? It is because I believe that these are closely tied to the questions we ask, and the interpretations we place upon our results, that I have chosen to examine research strategies. I will consider three such strategies, prefacing my remarks with the usual proviso that the boundaries between them cannot be sharply defined.

MODELS OF MAN

The least complicated strategy, the one that is easiest to explain and most difficult to defend in many behavioral studies, is to treat the nonhuman primate as a model of man. The best examples come from medical research. From where I stand — and it is admittedly very far from the center of the medical scene — the medical scientist working with primates can usually adopt an entirely matter-of-fact attitude toward his experimental subjects. He is concerned with a specific disease category. The disease is real, almost tangible; its symptoms are usually known; its annual cost can be reckoned in dollars, if not in human suffering; and the animal either contributes to the solution of the problem or it is abandoned in favor of a more promising candidate. The subject is selected because it is convenient, safe to work with, and a suitable model for the investigation of a specific disease category. Suitability requires only that the animal's primary response to the experimental condition be sufficiently close to the human to permit extrapolation to man. Beyond this relatively straightforward requirement, it makes no difference in how many other ways the animal differs from man. Indeed, we know that apart from the final validating test, a medical investigation may be carried out entirely on mice, rabbits or guinea pigs. The primates are required only when the disease fails to take in nonprimate forms.

The medical paradigm has had a definite impact on behavioral research, and it has been used in some investigations with notable success. Usually, however, the research can be characterized in terms of a limited and specifiable objective. For example, if we are going to study the effects of a new drug on memory, all we really require is a model that is capable of remembering; the question of how closely the model matches human performance in other ways is of small importance. Its sex life, its child-rearing practices, or, for that matter, even its general intelligence, need not concern us.

If we were to attempt to characterize this strategy in regard to its long-range objectives, we would probably come close to the truth by stating that it is a search for "common denominators," for effects that are sufficiently widespread or basic that they can be considered without reference to the idiosyncrasies of any particular species. Knowledge of such effects do, of course, have an important place in defining essential similarities, both within the primates and between the primates and other species.

But what happens when our interest shifts to questions in which species differences can no longer be disregarded? Suppose we are concerned with intellectual ability, or maternal behavior, or the effects of early experience on adult behavior? As the questions become more general, as more and more of the total organism becomes involved in the issue, the resemblance between the model and man becomes more tenuous, and the animal can no longer be regarded as a suitable substitute. When this occurs another research strategy, another way of looking at animal behavior, is required. Such a strategy was evolved early in the history of behavioral research and it has produced a distinguished record of achievement.

CONCEPTUAL MODELS

The essential feature of this second strategy is that it is concerned with the animal as a resource for constructing and testing conceptual models. The first strategy regards the animal as a model for man — as a stand-in to be used in situations where participation by human subjects is inconvenient or impossible — in contrast, the present strategy regards the model as something that the animal will help to create. The term *model*, of course, may not actually be used. More often, the scientist describes himself as investigating concept formation, or the mechanisms of hunger, or the influence of hormones on sexual performance. Although recognizing that behavior is a process in which the organism as a whole is essentially involved, he abstracts from the whole certain part-systems and studies them as though they were self-contained functional entities. In short, he is engaged in building a conceptual model of some behavioral mechanism or process.

Approaching behavior from this strategy frees the investigator from any obligation to claim that his results apply to man. His is a task of pure research, and he may even go so far as to disavow any interest in applying his findings or his theories to human behavior. There is, of course, no compelling reason why he should wish to do so. Indeed, if his research subject was not a monkey, or a chimpanzee, but a rat, or a fish, or a cockroach, we would probably all agree that his restraint was commendable.

But what if the subject is, in fact, a nonhuman primate? Does this change the situation in any way? It does, in that we can safely assume that the investigator in most instances is not completely indifferent to human problems; otherwise, he would not be doing research with primates, first, because monkeys and apes, being more costly and difficult to work with than other animals, are not subjects of choice unless the problem seems to demand them; and secondly, because the problems selected for study with monkeys and apes are often formulated in a way that betrays a fundamental interest in man. Monkeys and apes are being used to investigate such complex and peculiarly human issues as the nature of intelligence, the sources of neurotic behavior, and the origins of love. How far are we justified here in applying conceptual models developed on the nonhuman primates to the behavior of man? Although one naturally assumes that models derived from studies of monkeys and apes will be more relevant than those worked out on nonprimate forms, even this general assumption may be questionable in any specific instance. It now seems clear, for example, that in contrast to man, many primate species in nature restrict their mating activities to certain seasons of the year. Furthermore, the dorso-ventral mode of intercourse, and probably many of the finer details of sexual behavior, are closer to the basic mammalian pattern than to human norms.

The hard fact is — as we all know — that simple extrapolation from nonhuman conceptual models to the behavior of man is seldom possible. But if this is true, how do we appraise the relevance of our theories and findings on monkeys and apes to human behavior? How can we maximize the likelihood that our research efforts with the nonhuman primates will pay the richest possible dividends in terms of understanding man? These questions cannot be answered within the framework of the conceptual models research strategy. Insofar as it is concerned with maximizing relevance, this strategy appeals to face validity, to the resemblances between the phenomena observed in animal and in man. The difficulty, of course, is that the parallel may be trivial — perhaps less obviously so than if we were to equate the elaborate and colorful structures created by the bower bird with human artistic productions, or the periodic lethal migrations of the lemming with human mass suicide, but no less misleading for that reason.

The problem, as Howard F. Hunt has put it, is that "... without sensible attention to species differences in organization of behavior, development, and other biological arrangements, we will never be able to bridge the species gap inferentially, at will and with confidence." (3) The solution already hinted at in Hunt's statement is that the conceptual models strategy must be fitted within a broader approach. This approach, which represents the third strategy on my list, provides a different view of primate behavior, a perspective that I believe will permit us to realize most fully the potential contribution of the nonhuman primates to human behavior.

Evolutionary-comparative Perspective

As you may have anticipated, the perspective I refer to is evolutionary and comparative; it is a perspective that gives the differences among the primates as prominent a place as the similarities, and regards them as equally in need of explanation. While acknowledging the great diversity of behavioral adaptations among the living primates, it offers the hope that within that diversity patterns will be seen, that trends will be found yielding a fresh outlook on primate behavior and illuminating the special psychologic attributes of man.

These are more than wishful thoughts and pious hopes. Owing to the efforts of paleontologists, physical anthropologists, and other students of primate phylogeny, the basic framework for such an approach is available. One of its most important features is that it places the living primates within a graded series which approximates the actual sequence of primate evolution. (1, 7) Figure 1 illustrates the reconstruction of primate evolution presented by Simpson. (9) It is the customary phylogenetic tree with the various branches multiplying and diverging through time. Compare this with Figure 2, after Clark, which presents primate evolution in linear perspective. (1)

There is perhaps no need to emphasize that the living series only approximates the evolutionary series: We are not actually descended from the living apes, nor they from the monkeys as we know them today. What the linear model of primate evolution implies, however, is that the living chimpanzee is more representative of what we were in our remote past than of what we are today. It implies that we can use the living primates as a means for looking into our own past, for getting at the evolutionary trends that have culminated in human behavior.

You might reply that the approximation is crude, that the margin for error is so great that any "trends" derived from such a framework will be of doubtful utility. Rather than attempting a summary reply to such a charge, I

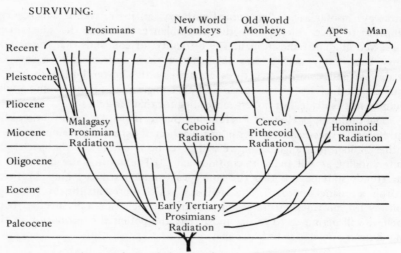

Fig. 1. Primate Phylogene, after Simpson, 1951.

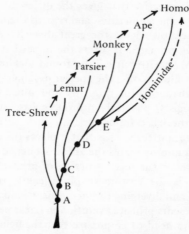

Fig. 2. Postulated Probable Linear Sequence of Hominid Ancestry. A-E: Hypothetical transitional stages at which different ramifications of the primates are presumed to have branched off. After Clark, 1959.

would like to consider one such trend which I have found especially instructive in thinking about problems of primate social development and socialization.

That man takes longer to grow up than most animals is common knowledge. It is less well known that the growth rate varies systematically among various species of primates. The lemurs grow more rapidly than the monkeys;

the monkeys more rapidly than the gibbon; and the gibbon is accelerated as compared to the three great apes and man. (8)

As one would expect, retardation in the rate of physical growth is accompanied by retardation in the rate of behavioral development: The rhesus monkey can walk unassisted by the age of 2 or 3 weeks; the chimpanzee, not before 4 months; and the human infant, usually not before the end of the first year of life. The same trend can be seen in other basic parameters of behavioral development.

One of the consequences of adopting an evolutionary-comparative perspective is that it engenders a concern for the adaptive value of species differences. It prompts us to ask, why this trend toward the prolongation of infancy and childhood? Why is it that the human organism takes so long to grow up? A common explanation is that prolonging immaturity gives the human being a longer time for learning. If we wish to be perverse and press the matter further, asking why the granting of more time for learning should have any special biologic utility, the answer usually given is that man has more time for learning because he has so much more to learn. At this point, the argument is often allowed to rest, since it is obvious to everyone that the growing child has a great deal to learn, particularly in a complex technologic society such as our own.

It seems to me, however, that this thesis overlooks one very important fact, namely, that the immature human is not a very efficient learner, at least on our standard measures of learned performance. Were it otherwise there would be no need to speak of readiness to learn, or to invent the concept of mental age. We may indeed need to learn more than any other primate, but the simple formula, "more time for learning is required, therefore immaturity is prolonged," appears to me basically unsound.

To be sure, any thesis that implicates learning in an evolutionary explanation of human growth is bound to have some appeal since *Homo sapiens* rightfully regards himself as the learning animal *par excellence*. But to make learning the focus of such an explanation is probably a mistake. We do, after all, possess many other distinctive traits. Together with the aforementioned slow rate of growth, we are the only primates whose habitual mode of locomotion is bipedal; whose bodies are not richly covered with hair; whose infants are totally incapable of clinging to their mothers and nursing unassisted; whose favored mode of coitus is ventro-ventral; whose infantile attachments persist throughout life.

Learning clearly cannot account for most of these attributes, but is there any other process that can? I believe that such a process was identified many years ago and was given the name neoteny. In some species it is possible for an individual to attain reproductive maturity while other biologic systems are still immature. The classic example of this phenomenon is the salamander, axolotl, an animal capable of reproduction while it is still in a larval state. This survival of juvenile traits into adulthood is what has been called neoteny.

It has been pointed out that in comparison with other primates, man can be considered neotenous. In such physical traits as brain weight, the feeble development of his brow ridges, the sparseness of hair over most of his body, the flatness of his face, and the small size of his canine teeth, the mature human resembles the young chimpanzee more than he does the adult animal. Similarly, the chimpanzee shows some indications of neoteny when compared to the monkey.

If physical structures are subject to neoteny, might not this also be true of behavior? Sir Gavin De Beer, in his fascinating book *Embryos and Ancestors*, concludes that this is a definite possibility. (2) The idea that the tendency to carry juvenile behavioral traits into reproductive maturity increases progressively in primate phylogeny offers some intriguing and, I believe, illuminating, implications for human conduct. Before we consider some of these, however, I propose to examine some rather specific features of primate infancy and childhood, which, if not actually the product of the same factors that result in neoteny, at least conform to the same phylogenetic trend. What I am referring to are species differences in the organization of the earliest integrated behavior patterns to appear in individual development.

The behavioral equipment of the newborn monkey, the newborn ape and the newborn human are so similar that they seem to have been fashioned on the same basic plan: All infants show the so-called Morro grasping response; all show the rooting, oral grasping, and sucking patterns involved in the first feeding reactions; all cry when things don't go their way; and all seem to be comforted by being properly held.

If we take a closer look at these primitive patterns, however, some important differences between species are at once apparent: Place a newborn rhesus monkey on its back and, unless it is given something to cling to, it immediately rights itself. Do the same thing with a newborn chimpanzee and it may perform weak and irregular sweeping movements with its arms, but it cannot right itself, and until it's 4 or 5 months old will not achieve any real facility in doing so. The same ability is not present in the human infant until it's about eight months old. Reflex grasping displays the same trend: An infant monkey can support itself with its hands alone for as long as thirty minutes, a chimpanzee for no longer than five minutes, and a human infant for two minutes or less. There is another difference between these species that is more difficult to quantify: Whatever it is that motivates the infant rhesus monkey to right itself, to suck and to cling, it is less prominent in chimpanzees and humans. For example, one receives the impression that the infant rhesus placed on its back is striving desperately to turn onto its belly; the chimpanzee in a supine position seems only moderately concerned, and so far as I know the human infant has never been described as showing a definite preference for either back or belly. Sucking is likewise more persistent, more efficient, more easily elicited, in the rhesus monkey than in either chimpanzee or human infant.

These differences reflect the greater developmental maturity of the newborn monkey, to be sure — but more important than this, they also indicate a fundamental difference between species in the way that infant behavior is organized. Viewed as a collection of responses preadapted to nursing and to maintaining contact with the mother, the behavior of the chimpanzee never displays the same reflex-like efficiency as that of the monkey. Indeed, if evolution had not brought about complementary changes in the behavior of the chimpanzee mother that permitted her to compensate for the behavioral deficiencies of her infant, its chances of survival would be slim. And no newborn human infant, of course, is able to cling, to find the nipple, and to nurse without extensive maternal assistance. The same infantile patterns that are highly organized in the rhesus monkey and have a life or death significance, often appear as fragmented, occasional responses in the human infant — as behavioral curios without obvious adaptive significance. I think you will agree in the light of these examples, that, even among the earliest organized behaviors to appear in primate infancy, there is evidence of a systematic trend, not only toward slowing down the rate of individual development, but also in the way that behavioral organization proceeds.

At later stages of behavioral development we find evidence of a similar trend toward the "loosening" of motor patterns. Consider locomotion: Rhesus monkeys progress chiefly by walking on all fours. Although they could doubtless be trained to walk bipedally on hands or feet they rarely do so spontaneously. The chimpanzee will often show elaborate variations on the basic quadrupedal pattern: They not only walk bipedally but may also progress by twirling like a ballerina, or in a series of somersaults, or by using their arms as though they were crutches, or by sliding on back or belly across the floor. Social relations, and the use of objects would furnish many other examples of a similar enrichment in motor patterns.

That a comparable change has occurred in motivational factors or the internal mechanisms that govern behavior seems most likely. In my own work with young chimpanzees I have found it necessary to speak in terms of a generalized motivational state, because more specific motives — fear, aggression, hunger, sex — do not seem capable of rendering a satisfactory account of their behavior. (5, 6) Could these changes — amounting to a generalized reduction in the strength, the efficiency, and the rigidity of instinctive patterns — be the result of the same evolutionary process that has caused the prolongation of primate infancy, and that has led to our description of man as the most neotenous of the primates? Robert Kuttner has argued, I think convincingly, that this is most likely, concluding that the enormous plasticity of behavior which is characteristic of the primates in general, and of man in particular, is the basis of our intellectual supremacy and the product of a neotenous trend. (4)

If it can be shown that behavioral neoteny is a generalized human characteristic, an important conceptual advance will have been made toward understanding many of the distinctive attributes of man: The loose relationship between hormonal factors and sexual and maternal activities; the prominence of such traits as playfulness, curiosity and inventiveness in the behavior of human adults would be seen in a somewhat different light. So, too, would the retention of other childlike elements in human behavior — the persistence of infantile needs and attachments, for example. Could neoteny be one of the reasons why it has been so difficult to demonstrate unequivocally the instinctive bases of human behavior, why the vagaries of individual experience seem to play so large a part in human destiny, and why the great students of human nature have felt compelled to postulate broad, pervasive forces such as the libido and the death wish to explain the instinctual basis of human conduct? Neoteny is probably a general primate characteristic, but man has gone farthest in this evolutionary venture — a venture in which the reliability and efficiency of instinctive patterns have been sacrificed to achieve the behavioral plasticity and the liberation of psychic energies that play so large a part in human frailties and human achievements.

CONCLUSION

But it is clear that I have moved quite beyond my stated intention of illustrating a research strategy into a discussion of a particular view of human nature and human evolution. Although such a digression can properly be regarded as indicating a lack of intellectual discipline in a speaker, it will have served a useful purpose if it has illustrated the utility and the intellectual appeal of an evolutionary-comparative approach. Such an approach is necessarily and essentially catholic and transdisciplinary; it feeds upon and nourishes many branches of science; and it will, I am convinced, play a critical part in determining the scope, and fulfilling the potential, of primate research.

REFERENCES

1. Clark, W.E. 1959. The Antecedents of Man. Edinburgh: Edinburgh University Press.
2. De Beer, G. 1958. Embryos and Ancestors, 3rd ed. London: Oxford University Press.
3. Hunt, H.F. 1964. Problems in the interpretation of "experimental neurosis." Psychol. Rep. 15:27-35.
4. Kuttner, R. 1960. An hypothesis on the evolution of intelligence. Psychol. Rep. 6:283-89.
5. Mason, W.A. 1965. Determinants of social behavior in young chimpanzees. In Behavior of Non-human Primates, eds. A.M. Schrier, H.F. Harlow and F. Stollnitz, pp. 335-64. New York: Academic Press.
6. ____. 1967. Motivational aspects of social responsiveness in young chimpanzees. In Early Behavior: Comparative and Developmental Approaches, ed. Harold W. Stevenson, pp. 103-126. New York: John Wiley & Sons.
7. Napier, J. 1962. The evolution of the hand. Sci. Amer. 207:56-61.
8. Schultz, A.H. 1956. Postembryonic age changes. In Primatologia I, eds. H. Hofer, A.M. Schultz and D. Starck, pp. 887-959. Basle: S. Karger.
9. Simpson, G.G. 1951. The Meaning of Evolution. New York: New American Library.

Appendix

The papers which make up this reader are of a general theoretical nature, examples of alternative approaches to the analysis and interpretation of primary data gathered over the past decade and a half. Works cited by their authors provide an obvious first resource for further study of nonhuman primate behavior; however, students may find the short list of journals, monographs, and articles which follows a more convenient starting point. Some of the articles listed update information presented in one or the other of the foregoing papers, which were included in this collection rather for their representativeness of current theory in development than for up-to-the-minute accuracy of factual detail.

Resources for the Study of Nonhuman Primate Behavior

Major Journals Devoted to Primates

Folia Primatologica
Primates

Some Journals in Which Reports of Primate Behavior Frequently Appear

American Anthropologist
American Journal of Physical Anthropology
Animal Behaviour
Behaviour, an International Journal of Comparative Ethology
Journal of Comparative and Physiological Psychology
Journal of Mammalogy
Man
Nature
Proceedings (and Symposia) of the Zoological Society of London
(after vol. 145, *see* Journal of Zoology)
Psychological Reports
Science

Recent Monographs: Representative Detailed Reports of Long-term Field Studies

Altmann, Stuart A., and Jeanne Altmann: Baboon Ecology, African Field Research (The University of Chicago Press, Chicago and London 1970).
Jolly, Alison: Lemur Behavior, A Madagascar Field Study (The University of Chicago Press, Chicago and London 1966).
Kummer, Hans: Social Organization of Hamadryas Baboons, a Field Study (The University of Chicago Press, Chicago and London 1968).
Poirier, Frank E.: The Nilgiri Langur of South India; in L.A. Rosenbloom, ed., Primate Behavior, Developments in Field and Laboratory Research, Vol. 1, pp. 251-383 (Academic Press, New York and London 1970).

Schaller, George B.: The Mountain Gorilla, Ecology and Behavior (The University of Chicago Press, Chicago and London 1963).
van Lawick-Goodall, Jane: The Behavior of Free-living Chimpanzees in the Gombe Stream Reserve. Anim. Behav. Monog. *1*:161-311 (1968).

SEE ALSO

DeVore, Irven (ed.): Primate Behavior, Field Studies of Monkeys and Apes (Holt, Rinehart and Winston, New York and London 1965).
Carpenter, C. Ray: Naturalistic Behavior of Nonhuman Primates (The Pennsylvania State University Press, University Park, Pennsylvania 1964).

A SAMPLING OF SHORTER, PROBLEM-ORIENTED
PAPERS & REVIEW ARTICLES

Altmann, S.A.: Sociobiology of Rhesus Monkeys IV: Testing Mason's Hypothesis of Sex Differences in Affective Behavior. Behaviour *32*:49-69 (1968).
Anthoney, T.R.: The Ontogeny of Greeting, Grooming and Sexual Motor Patterns in Captive Baboons (Superspecies: *Papio cynocephalus*). Behaviour *31*:358-372 (1968).
Bates, B.C.: Territorial Behavior in Primates: A Review of Recent Field Studies. Primates *11*:271-284 (1970).
Bernstein, I.S.: Primate Status Hierarchies; in L.A. Rosenbloom, ed., Primate Behavior, Developments in Field and Laboratory Research, Vol. 1, pp. 71-109 (Academic Press, New York and London 1970).
Crook, J.H.: The Socio-ecology of Primates; in J.H. Crook, ed., Social Behavior in Birds and Mammals (Academic Press, New York and London 1970).
Fedigan, L.M.: Roles and Activities of Male Geladas (*Theropithecus gelada*). Behaviour *41*:82-90 (1972).
Furuya, Y.: On the Fission of Troops of Japanese Monkeys I. Five Fissions and Social Changes between 1955 and 1966 in the Gagyusan Troop. Primates *9*:323-350 (1968).
Furuya, Y.: On the Fission of Troops of Japanese Monkeys II. General View of Troop Fission of Japanese Monkeys. Primates *10*:47-69 (1969).
Harlow, H.F., M.K. Harlow and S.J. Suomi: From Thought to Therapy: Lessons from a Primate Laboratory. American Scientist *59*:538-549 (1971).
Kaufman, J.H.: Behavior of Infant Rhesus Monkeys and Their Mothers in a Free-ranging Band. Zoologica *51*:17-32 (1966).
Koford, C.B.: Rank of Mothers and Sons in Bands of Rhesus Monkeys. Science *141*:356-357 (1966).
Koyama, N.: Changes in Dominance Rank and Division of a Wild Japanese Monkey Troop in Arashiyama. Primates *11*:335-390 (1970).
Kummer, H.: Tripartite Relations in Hamadryas Baboons; in S.A. Altmann,

ed., Social Communication among Primates (University of Chicago Press, Chicago and London 1967).

Lee, R.B.: What Hunters Do for a Living, or How to Make Out on Scarce Resources; in R.B. Lee and I. DeVore, eds., Man the Hunter (Aldine, Chicago 1968).

Lindburg, D.G.: Rhesus Monkeys: Mating Season Mobility of Adult Males. Science 166:1176-1178 (1969).

Mitchell, G.: Paternalistic Behavior in Primates. Psychological Bulletins 71:229-417 (1969).

Nishida, T.: A Sociological Study of Solitary Male Monkeys. Primates 7:141-204 (1966).

Rowell, T.E., A.D. Nasar and A. Omar: A Quantitative Comparison of a Wild and a Caged Baboon Group. Animal Behaviour 15:499-501 (1967).

Rowell, T.E.: The Social Development of Baboons in Their First Three Months. Journal of Zoology, Lond. 155:461-483 (1968).

Rowell, T.E.: Long-Term Changes in a Population of Ugandan Baboons. Folia Primatologica 11:241-254 (1969).

Sade, D.S.: Some Aspects of Parent-Offspring and Sibling Relations in a Group of Rhesus Monkeys, with a Discussion of Grooming. Am. J. phys. Anthrop. 23:1-17 (1965).

Struhsaker, T.T.: Correlates of Ecology and Social Organization among African Cercopithecines. Folia Primatologica 11:80-118 (1969).

Suzuki, A.: An Ecological Study of Chimpanzees in a Savanna Woodland. Primates 10:103-148 (1969).

Menzel, E.W., Jr.: Communication about the Environment in a Group of Young Chimpanzees. Folia Primatologica 15:220-232 (1971).

Vandenbergh, J.G.: The Development of Social Structure in Free-ranging Rhesus Monkeys. Behaviour 29:179-194 (1967).

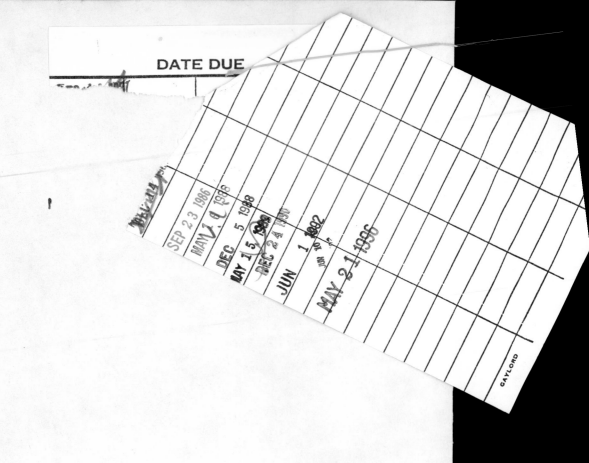